Trees: Propagation and Conservation

Ankita Varshney • Mohammad Anis

Trees: Propagation and Conservation

Biotechnological Approaches for Propagation of a Multipurpose Tree, *Balanites aegyptiaca* Del.

Dr. Ankita Varshney
Department of Botany
Plant Biotechnology Laboratory
Aligarh Muslim University
Aligarh
India

Prof. Mohammad Anis
Department of Botany
Plant Biotechnology Laboratory
Aligarh Muslim University
Aligarh
India

ISBN 978-81-322-1700-8   ISBN 978-81-322-1701-5 (eBook)
DOI 10.1007/978-81-322-1701-5
Springer New Delhi Dordrecht Heidelberg London New York

Library of Congress Control Number: 2013957489

© Springer India 2014
This work is subject to copyright. All rights are reserved by the Publisher, whether the whole or part of the material is concerned, specifically the rights of translation, reprinting, reuse of illustrations, recitation, broadcasting, reproduction on microfilms or in any other physical way, and transmission or information storage and retrieval, electronic adaptation, computer software, or by similar or dissimilar methodology now known or hereafter developed. Exempted from this legal reservation are brief excerpts in connection with reviews or scholarly analysis or material supplied specifically for the purpose of being entered and executed on a computer system, for exclusive use by the purchaser of the work. Duplication of this publication or parts thereof is permitted only under the provisions of the Copyright Law of the Publisher's location, in its current version, and permission for use must always be obtained from Springer. Permissions for use may be obtained through RightsLink at the Copyright Clearance Center. Violations are liable to prosecution under the respective Copyright Law.
The use of general descriptive names, registered names, trademarks, service marks, etc. in this publication does not imply, even in the absence of a specific statement, that such names are exempt from the relevant protective laws and regulations and therefore free for general use.
While the advice and information in this book are believed to be true and accurate at the date of publication, neither the authors nor the editors nor the publisher can accept any legal responsibility for any errors or omissions that may be made. The publisher makes no warranty, express or implied, with respect to the material contained herein.

Printed on acid-free paper

Springer is part of Springer Science+Business Media (www.springer.com)

Preface

Micropropagation or plant tissue culture comprises a set of in vitro techniques, methods, and strategies that are part of the group of technologies called plant biotechnology. Plant tissue culture technology is playing an increasingly important role in basic and applied studies. Also, the application of tissue-culture technology, as a central tool or as an adjunct to other methods, including recombinant DNA techniques, is at the vanguard in plant modification and improvement for agriculture, horticulture and forestry.

Tree propagation in vitro has been a difficult proposition compared to other plants. The response of explants is primarily determined by genotype, physiological state of the tissue, and time of the year when the explants are collected and cultured. While most nurserymen have been introduced to the techniques and advantages of micropropagation, few have ventured to use it as a propagation tool. The applicability of micropropagation for woody trees has been demonstrated as feasible since all aspects of the technology have confirmed the fact that trees produced by this method look like and grow like their counterparts produced by traditional methods of cloning. The tree species in forest, plantation, and urban environments are important biological resources that play a major role in the economy and the ecology of terrestrial ecosystems, and they have aesthetic and spiritual value. Because of these many values of the tree species, preserving forest tree biodiversity through the use of biotechnological approaches should be an integral component in any forestry program in addition to large-scale ecologically sustainable forest management and preservation of the urban forest environment. Biotechnological tools are available for conserving tree species as well as genetic characterization that will be needed for deployment of germplasm through restoration activities. This book concentrates on the biotechnological tools available for multiplying, conserving, characterizing, evaluating, and enhancing the forest tree biodiversity, especially a multipurpose semi-arid tree *Balanites aegyptiaca* Del.

Micropropagation has gained the status of a multibillion dollar-industry being practiced in hundreds of biotechnology laboratories and nurseries throughout the world. As compared to annual herbaceous plants, tissue culture of trees is a difficult proposition mainly because of their being intractable to regenerate and slow growing and having the problems of dormancy, juvenility versus maturity, phenolics, and endogenous infection as well as great physiological variation in explants collected from fields besides less consistent efforts made at the global level. In vitro micropropagation has proved, in the recent past, a means for supply of planting material for forestry. Different basal media, plant growth regulators, media additives, and carbohydrate sources are being used to manipulate culture conditions in vitro for propagation of forest trees. Micropropagation of forest trees in vitro is not only a means for mass scale propagation of superior clones of tree species but it can also be used for developing transgenic plants and conservation of germplasm through cryopreservation.

We anticipate that the conclusion of this book will be an increased awareness of the fact that there is still a great need for strategic research in applied sciences like plant biotechnology. The contents of the book also provide an indication of some of the information in which plant biotechnology (in vitro culture of trees) is likely to go in the coming years. At the very least, we feel that it should provide a source of background information and references to both students and researchers alike who wish to initiate or broaden their interests in the field.

It is noticeable that many of the chapters described in this book tend to be complimentary to each other, particularly where a combination of techniques may be required to achieve an ideal objective. This reinforce a vision held by us that plant biotechnology can rightly be considered a novel and key area of research involving both the applied and basic aspects of plant sciences.

September 2013

Dr. Ankita Varshney
Prof. Mohammad Anis

Acknowledgments

The research support and assistance rendered by the Department of Science and Technology and the University Grant Commission, Govt. of India, New Delhi, in the form of DST-FIST (2011–2016) and UGC-DRS-I (2009–2014) programs is duly acknowledged. The award of Young Scientist under DST-FAST TRACK scheme to Ankita Varshney is also gratefully acknowledged. We would also like to acknowledge the help rendered by Dr. Anushi Arjumend Jahan, in preparing the layout of the book. We would also like to place on record our sincere thanks to other colleagues in Plant Biotechnology Laboratory, Department of Botany, Aligarh Muslim University, Aligarh, for their support and cooperation. We would also like to thank Springer for providing us the opportunity to publish this book.

Contents

1 Introduction 1
   1.1 Global and National Scenarios 1
   1.2 In Vitro Approaches 3
   1.3 Constraints with Tree Tissue Culture 4
   1.4 Plant Tissue Culture for Trees 4
   1.5 Taxonomy 5
        1.5.1 Scientific Classification 5
        1.5.2 Binomial name 5
        1.5.3 Common names 5
        1.5.4 Habitat 5
        1.5.5 Morphological Description 5
        1.5.6 Active Constituents 6
        1.5.7 Medicinal Properties and Uses 7
        1.5.8 Other Uses 7
        1.5.9 Conventional Propagation Methods
               and its Limitations 7
   References 8

2 Review of Literature 11
   2.1 Introduction 11
   2.2 Micropropagation 13
   2.3 Various Approaches for Micropropagation 14
        2.3.1 The Propagation of Plants from Axillary
               Buds or Shoots 14
        2.3.2 Propagation by Adventitious Shoot Organogenesis 15
   2.4 Factors Affecting In Vitro Shoot Regeneration
        and Growth of Plants 17
        2.4.1 Explant Type 17
        2.4.2 Plant Growth Regulators 18
        2.4.3 Medium pH Levels 21
        2.4.4 Basal Media 22
        2.4.5 Carbohydrate Source 24
        2.4.6 Subculture Passages 25
   2.5 Rooting of In Vitro-Regenerated Shoots 26
   2.6 Acclimatization and Hardening of Plantlets 27
   2.7 Advancement in Plant Tissue Culture: Synthetic
        Seed Technology 28

	2.8	Clonal Fidelity of Micropropagated Plants	29
		2.8.1 PCR-Based DNA Markers	30
	2.9	Antioxidant Enzymes	31
	References		33
3	**Materials and Methods**		49
	3.1	Plant Materials and Explant Source	49
	3.2	Surface Disinfection of Seeds and Explants	49
	3.3	Establishment of Aseptic Seedlings and Preparation of Explants	49
	3.4	Culture Media	50
		3.4.1 Composition of Basal Media	50
		3.4.2 Preparation of Stock Solutions	50
	3.5	Plant Growth Regulators	51
	3.6	Adjustment of pH, Gelling of Medium, Carbon Source and Sterilization	52
	3.7	Sterilization of Glassware and Instruments	52
	3.8	Sterilization of Laminar AirFlow Hood	52
	3.9	Inoculation and Culture Conditions	52
	3.10	Rooting	52
	3.11	Acclimatization and Hardening of Plantlets	52
	3.12	Synthetic Seed	53
		3.12.1 Explant Source	53
		3.12.2 Encapsulation Matrix	53
	3.13	Physiological and Biochemical Studies of in vitro Regenerated Plants During Acclimatization	53
		3.13.1 Chlorophyll and Carotenoid Estimation	53
	3.14	Assessment of Antioxidant Enzyme Activities	54
		3.14.1 Superoxide Dismutase	54
		3.14.2 Catalase	55
		3.14.3 Ascorbate Peroxidase	56
		3.14.4 Glutathione Reductase	56
		3.14.5 Soluble Protein	57
	3.15	Anatomical Studies	58
		3.15.1 Fixation and Storage	58
		3.15.2 Embedding, Sectioning and Staining	58
	3.16	Chemicals and Glassware Used	58
	3.17	Statistical Analysis	58
	3.18	Genomic DNA Isolation and Purification	58
		3.18.1 Preparation of Stock Solutions Required for DNA Extraction	58
		3.18.2 Extraction and Purification Protocol	59
		3.18.3 Quantitative and Qualitative Assessment of Genomic DNA	59
	3.19	PCR Amplification of DNA Using ISSR Primers	60
		3.19.1 Polymerase Chain Reaction Amplification	60
		3.19.2 ISSR-PCR with Genomic DNA	61

		3.19.3	Analysis of PCR Products by Agarose Gel Electrophoresis	61
		3.19.4	Data Scoring and Analysis	61
	References			61
4	**Results**			**63**
	4.1	Direct Shoot Regeneration		63
		4.1.1	Establishment of Aseptic Seedling	63
		4.1.2	Regeneration from CN Explant Excised from 15-day-old Aseptic Seedlings	63
		4.1.3	Regeneration from Nodal Explant Excised from 4-week old Aseptic Seedlings	68
		4.1.4	Regeneration from Root Explants Excised from 4-week-old Aseptic Seedlings	75
		4.1.5	Induction of Multiple Shoots from Intact Seedlings	80
		4.1.6	Regeneration from Mature Nodal Explants Excised from 10-year-old Candidate Plus Tree	83
		4.1.7	Rooting of Microshoots	88
		4.1.8	Acclimatization	91
		4.1.9	Synthetic Seeds	91
		4.1.10	Rooting in Synthetic Seeds and Establishment of Plants in Soil	92
	4.2	Assessment of Physiological and Biochemical Parameters		93
		4.2.1	Photosynthetic Pigments	93
		4.2.2	Antioxidant Enzyme Activities	93
	4.3	Clonal Fidelity in TC-Raised Plantlets Derived from Mature Nodal Explants		93
5	**Discussion**			**101**
	5.1	Effect of Plant Growth Regulators on Different Explants		101
	5.2	Effect of Different Media, Sucrose Concentrations and pH Levels		104
	5.3	Rooting and Acclimatization		104
	5.4	Synthetic Seeds		106
	5.5	Antioxidant Enzymes Activities		107
	5.6	Assessment of Genetic Fidelity		108
	References			108
6	**Summary and Conclusions**			**115**
	6.1	Summary		115
	6.2	Conclusions		116

About the Authors

Ankita Varshney obtained her Ph.D. degree in Botany (Plant Biotechnology) in 2012 from the Aligarh Muslim University, Aligarh. Currently, she is working as a Young Scientist under DST-FAST TRACK scheme. Earlier, she worked as Research Associate of CSIR, New Delhi. She has published 10 research papers in journals of national and international repute. Her research focuses on in vitro regeneration, biochemical and molecular characterization of medicinal plants.

Mohammad Anis is a Professor and a Former Chairman in the Department of Botany at the Aligarh Muslim University, Aligarh. He has 26 years of teaching and research experience in cytogenetics, plant tissue culture and morphogenesis. He has published more than 150 research papers in journals of national and international repute. His current research focuses on morphogenesis and in vitro propagation of medicinal and woody plant species. He has been a Programme Coordinator of DST-FIST, DBT-HRD and UGC-DRS-I projects in the department. He has been conferred with "Vigyan Ratan Samman-2010" by the U.P. Council of Science and Technology.

Abbreviations

2-iP	2-isopentenyladenine
APX	Ascorbate peroxidase
B_5	Gamborg's medium
%	Percent
°C	Degree celsius
2,4-D	2,4-Dichlorophenoxyacetic acid
BA	6-Benzyladenine
BSA	Bovine serum albumin
CAT	Catalase
Chl	Chlorophyll
cm	Centimeter
CN	Cotyledonary node
CTAB	Cetyl-trimethylammonium bromide
d	Days
dATP	Dinucleotide adenine triphosphate
dCTP	Dinucleotide cytosine triphosphate
DDW	Double distilled water
dGTP	Dinucleotide guanine triphosphate
DNA	Deoxyribonucleic acid
dNTPs	Dinucleotide triphosphates
dTTP	Dinucleotide thiamine triphosphate
ε	Extinction coefficient
EDTA	Ethylene diamine tetraacetic acid
EU	Enzyme units
FAA	Formalin acetic acid
FW	Fresh weight
g	Gram
GR	Glutathione reductase
GSH	Reduced glutathione
GSSG	Oxidized glutathione
h	Hour
H_2O_2	Hydrogen peroxide
HCl	Hydrochloric acid
IAA	Indole-3-acetic acid
IBA	Indole-3-butyric acid
ISSR	Inter-sequence simple repeats
Kn	Kinetin (6-furfurylaminopurine)

L_2	Phillips and Collin's medium
M	Molarity
Mg	Miligram
mg/l	Miligram per liter
$MgCl_2$	Magnesium chloride
min	Minute
ml	Milliliter
mm	Millimeter
mM	Milimolar
MS	Murashige and Skoog medium
N	Normality
NAA	α-naphthalene acetic acid
NaCl	Sodium chloride
NADPH	Nicotinamide adenine dinucleotide phosphate
NaOH	Sodium hydroxide
NBT	Nitroblue tetrazolium
nm	Nanometer
OD	Optical density
PCR	Polymerase chain reaction
PGRs	Plant growth regulators
PPFD	Photosynthetic photon flux density
ppm	Parts per million
PVP	Polyvinyl pyrrolidone
RAPD	Random amplified polymorphism DNA
RH	Relative humidity
RNA	Ribonucleic acid
ROS	Reactive oxygen species
rpm	Rotation per minute
sec	Second
SOD	Superoxide dismutase
TBE	Tris-borate EDTA
TCA	Trichloroacetic acid
TDZ	Thidiazuron
TE	Tris-HCl EDTA
U	Unit
UBC	University of British Columbia
UV	Ultraviolet
V	Volt
v/v	Volume by volume
W	Watt
w/v	Weight by volume
μg	Microgram
μM	Micromolar

Introduction

1

Abstract

Biodiversity, as this assemblage of life-forms is referred to, has now been acknowledged as the foundation for sustainable livelihood and food security. Forests are one of the most valuable ecosystems in the world, containing more than 60% of the world's biodiversity. Forest trees are recognized as a raw-material base for industrial and domestic wood products, which perpetually provide renewable energy, fiber, and timber. Besides other valuable products, several trees are recognized for their medicinal and pharmaceutical importance. The economic benefits of planted forests have led to their widespread adoption throughout the world. To maintain and sustain forest vegetation, conventional approaches have been exploited for propagation and improvement, but tree-breeding efforts are restricted to the most valuable and fast-growing species. However, such methods are limited by several inherent bottlenecks because trees are generally slow-growing, long-lived, sexually self-incompatible, and highly heterozygous plants. Tissue culture and other biotechnological approaches offer tremendous scope towards the desired objectives. This chapter deals with a brief introduction about global and national status of forests and applications and limitations of plant tissue culture for trees with a special reference to a semiarid tree, *Balanites aegyptiaca* (L.) Del.

1.1 Global and National Scenarios

The planet Earth is endowed with rich varieties of life-forms, and the teeming millions of these living organisms have been well knit by the laws of nature. The interdependence of various life-forms starting from the unicellular primary producers to the complexly built higher plants and animals is a unique feature of this green planet. Biodiversity, as this assemblage of life-forms is referred to, has now been acknowledged as the foundation for sustainable livelihood and food security. Also, scientists have estimated that more than 50 million species of plants and animals, including invertebrates and microorganisms, exist on earth and hardly 2 million of them have been described so far.

Forests are one of the most valuable ecosystems in the world, containing more than 60% of the world's biodiversity. This biodiversity has multiple social and economic values, apart from its intrinsic value, varying from the important ecological functions of forests in terms of soil and watershed protection to the economic value of the numerous products which can be extracted from the forest. For many indigenous and other forest-dependent people, forests are their livelihood. They provide them with edible and medicinal plants, bushmeat, fruits, honey, shelter, firewood, and many other goods, as well as with cultural and spiritual values.

On a global scale, all forests play a crucial role in climate regulation and constitute one of the major carbon sinks on earth, their survival thus preventing an increase in the greenhouse effect. Forest trees are recognized as a raw-material base for industrial and domestic wood products, which perpetually provide renewable energy, fiber, and timber. The economic benefits of planted forests have led to their widespread adoption throughout the world. Globally, 48 % of the forest plantation is established for industrial use, 26 % for nonindustrial use (fuelwood, soil, and water conservation), while the remaining 26 % is not specified. The current extent of world's plantation forest area is about 187 million hectares (m ha) with the annual planting of 4.5 m ha (FRA 2001).

India lies on the tropic of cancer and is a known mega biodiversity center with 8 % of the global biodiversity in 2.4 % land. The country is ranked tenth among the plant-rich countries of the world and fourth among the countries of Asia. It poses out 20,000 species of higher plants, one third of it being endemic and 500 species are categorized to have medicinal values (Krishnan et al. 2011). In addition, it is one of the largest hardwood plantation resources of about 32.5 m ha with *Eucalyptus, Acacia,* and teak as the major species (FRA 2001), whereas bamboo species occupy 8.96 m ha. The annual planting target of India is about 3 m ha, which requires roughly 6,100 million seedlings for ten major species, viz., *Eucalyptus,* bamboos, *Acacia, Albizia, Prosopis cineraria, P. juliflora, Casuarina, Dalbergia,* conifers, and teak (Gurumurthi 1994).

The National Forestry Action Programme (NFAP) and National Forest Policy of India have identified the expansion of area under forest plantations with increased productivity as one of the important thrust areas in forestry. The significance of plantation establishment and the use of improved planting stock have also been emphasized in a number of recent reviews (Kanonshi 1997; Evans 1998; Pandey and Ball 1998; Yasodha et al. 2004; Anis et al. 2012).

Besides other valuable products, several trees are recognized for their medicinal and pharmaceutical importance. Today, at a local level, an extremely wide range of plant species is used for medicinal purposes, the World Health Organization (WHO) has listed more than 21,000 plant names (including synonyms) that have reported medicinal uses around the world. Thousands of medicinal and aromatic plants have already been reported in Indian literature pertaining to their medicinal importance. In this context, India has recognized just more than 7,500 plant species having true medicinal values, but more than 500 traditional communities use about 800 plant species for curing different diseases (Kamboj 2000).

In developing countries as a whole, it is estimated that more than 6,000 plants are utilized in traditional medicines (Srivastava et al. 1995). In India, about 33 % plant species are trees while about 52 % are shrubs and herbs. About 1,500 plant species are used for ethical and classical formulations and home remedies based on the Indian System of Medicines (ISMs) such as Ayurveda, Siddha, and Unani. It is estimated that Indian consumption alone of these medicinal plants (188 tonnes) is used for culinary purposes and about 12 tonnes are consumed for medicinal and cosmetic preparations. In the last century, roughly 121 pharmaceutical products were formulated based on the traditional knowledge obtained from various sources. Currently, 80 % of the world population depends on plant-derived medicine for the first line of primary health care for human alleviation because it has no side effects (Vines 2004).

Worldwide, many plant species are threatened with extinction because of the gradual disappearance of the terrestrial natural ecosystem caused by various human activities. Often, this is due to the clearing of indigenous vegetation for agriculture and the resulting erosion, salinization, and invasion by alien species, but more recently climate change is looming as a significant new threat. More than 50 % of the world plant species are endemic to 34 global biodiversity hot spots which once covered 15.7% of the earth's land surface and which are now reduced to 2.3%. The International Union for Conservation of Nature Red List of Threatened Plants, first published in 1998 (IUCN 1998), lists more than 8,000 species currently in danger (IUCN 2010). There are numerous estimates of predicted extinction rates but the consensus view is that 15–20% of all plant species could become extinct by 2020.

Besides this, demographic changes, the growing size of population of the world, and increasing urbanization have had and will continue to have a major impact on forest cover and condition, demand for wood and nonwood forest products, and the ability of forests to fulfill essential environmental functions. Political and economic trends affecting the forestry sector are decentralization, privatization, trade liberalization, and globalization of world economy, and overall economic growth coupled with a widening gap between the rich and the poor in many countries.

Deforestation and forest degradation are occurring in drylands and upland areas. These already have limited forest cover and are fragile environments susceptible to soil erosion and other forms of degradation. In these areas, the communities are highly dependent on forests for food, fuel, and income. The most important direct causes of deforestation include logging, conversion of forested lands for agriculture and cattle-raising, urbanization, mining and oil exploitation, acid rain, and fire. However, there has been a tendency of highlighting small-scale migratory farmers or "poverty" as the major cause of forest loss. Such farmers tend to settle along roads through the forest, to clear a patch of land and to use it for growing subsistence or cash crops. In tropical forests, such practices tend to lead to rapid soil degradation as most soils are too poor to sustain agriculture. Consequently, the farmer is forced to clear another patch of forest after a few years. The degraded agricultural land is often used for a few years more for cattle grazing. This is a death sentence for the soil, as cattle remove the last scarce traces of fertility. The result is an entirely degraded piece of land which will be unable to recover its original biomass for many years.

To maintain and sustain forest vegetation, conventional approaches have been exploited for propagation and improvement, but tree-breeding efforts are restricted to the most valuable and fast-growing species. The major challenge for the tree breeder is to quickly transfer the ever-improving material from the breeding programs into plantations. The basic method of achieving such transfers is the use of seeds from orchards or clonal propagation. However, such methods are limited with several inherent bottlenecks because trees are generally slow-growing, long-lived, sexually self-incompatible, and highly heterozygous plants. Due to the prevalence of high heterozygosity in these species, a number of recessive deleterious alleles are retained within populations, resulting in high genetic load and inbreeding methods such as selfing and backcrossing, and makes it difficult to fix desirable alleles in a particular genetic background (Williams and Savolainen 1996). Thus, conventional breeding is rather slow and less productive and cannot be used efficiently for the genetic improvement of trees.

Tissue culture and other biotechnological approaches offer tremendous scope towards the desired objectives. Exploration, collection, characterization, evaluation, domestication/cultivation, and ex situ conservation in gene banks eventually support their sustained use by supplying good quality planting materials and certified raw drugs.

1.2 In Vitro Approaches

The worldwide importance of forestry, summed to the lengthy generation cycles of tree species, makes the development of new technologies unavoidable that complement conventional tree-breeding programs in order to obtain improved genotypes. In this regard, a new set of tools has become available in the past 20 years that combined with traditional plant breeding will allow people to generate products, the genetically improved varieties of the future. This set of tools comes under the general title of "Biotechnology." Three specific biotechnological tools have been successfully used in several programs of plant conservation, namely: (1) tissue culture techniques for in vitro propagation, (2) the use of molecular marker to assess the degree of variability among population, and (3) techniques of long-term conservation such as encapsulation and cryopreservation.

Plant tissue culture techniques are particularly relevant, because they not only become an alternative for large-scale propagation of individuals that are threatened, reduce production costs, and increase gains to the industry, but also provide ecological advantages as in phytoremediation or in the establishment of artificial plantings in weed-infested sites. The use of more efficient trees may reduce the soil areas required to produce the goods and services derived from tree plantations, thus reducing the use of natural ecosystems for sustainable and agricultural purposes. In India, the major accomplishment has been made in the in vitro propagation of various plant species, including forest trees like teak, *Eucalyptus,* bamboos, sandal, *Vitex negundo, Albizia lebbeck,* and *Tecomella undulata* (Sharma 2002; Anis et al. 2012).

1.3 Constraints with Tree Tissue Culture

Tissue culture of tree species often poses problems which are either absent or of lesser significance when culturing herbaceous species. Tree species have been cultured in vitro since the 1930s. Since that time, much progress has been made in the culture of tissues, organs, cells, and protoplasts of tree species. That is not to say that all woody plants, including forest trees, can be induced to grow and differentiate in vitro. Some do, while others are still recalcitrant. In general, juvenile tissues from trees are more responsive to in vitro manipulations than the mature tissues. Trees have long generation cycles and have an extended vegetative phase, ranging from 10 to 50 years. During the juvenile phase, starting from the embryo and perhaps lasting up to a decade, tissues from tree species are responsive to in vitro conditions. As maturation sets in, tissues from mature trees become less responsive in tissue culture. Other than the age of a tree, the response of an explant/tissue is also determined by the genotype, physiological state of the tissue, time of the year when the explant is cultured, and the composition of medium. Also, dependence on the mature or field-grown materials poses surface sterilization problems (Pierik 1997).

Even with these limitations, tissue culture has been successfully applied to a wide range of trees and its implication has been mentioned in some exhaustive review of Thorpe and Harry (1991), Harry and Thorpe (1994), McCown (2000), Giri et al. (2004), Yasodha et al. (2004), and Anis et al. (2012).

1.4 Plant Tissue Culture for Trees

Since the beginning of domestication and cultivation of plants, human beings are looking for techniques that can help them to produce maximum number of individuals from the minimum number or quantity of propagules. Plant tissue culture is the ultimate finding of their inquiry towards mass multiplication of plants using minimum quantity of propagules. Tissue culture refers to the aseptic growth of cells, tissues, or organs in artificial media. Tissue culture approaches to conservation are appealing as the procedure ensures that propagation is possible for species where:
1. Source explants are in limited abundance.
2. Inbreeding has resulted in lowered seed yields (Hendrix and Kyhl 2000).
3. Species possess complex, often unresolved seed dormancy mechanisms (Merritt and Dixon 2003; Merritt et al. 2007).
4. Seed availability is hampered by low and erratic seed set and viability due to environmental stresses such as drought, predation, or disease (or simply a paucity of information on optimal time for seed collection).

Some of the advantages of this technique are that heterozygous materials may be perpetuated without much alteration; it is easier and faster; the dormancy problem is eliminated; and juvenile stage is reduced. It is also a means for perpetuating clones that do not produce viable seeds or that do not produce seeds at all. It is now a well-established technology. Like many other technologies, it has gone through different stages of evolution: scientific curiosity, research tool, novel applications, and mass exploitation.

Today plant tissue culture applications encompass much more than clonal propagation and micropropagation. The range of routine technologies has expanded to include somatic embryogenesis, somatic hybridization, and virus elimination as well as the application of bioreactors to mass propagation. These techniques, particularly micropropagation methods, are currently in use for the large-scale multiplication of important tree species; woody biomass production; production of valuable secondary metabolites including pharmaceuticals, pigments, and other chemicals; and conservation of elite and rare germplasms (Giri et al. 2004).

In general, three modes of in vitro plant regeneration have been in practice, organogenesis, embryogenesis, and axillary proliferation. The difference mainly matters when it relates to the genetic stability of the resulting micropropagated plants; the obvious option then would be axillary and adventitious shoot proliferation. In vitro micropropagation has proved to be a means for supply of planting material for forestry in the recent past (Ahuja 1993; Lakshmisita and Raghavaswamy 1998).

In the present book, an important forest tree species namely, *Balanites aegyptiaca* (L.) Del. has been selected for its large-scale propagation which can be used for reintroduction and conservation purposes through biotechnological approaches.

1.5 Taxonomy

1.5.1 Scientific Classification

Kingdom: Plantae
Division: Magnoliophyta
Class: Magnoliopsida
Order: Zygophyllales
Family: Zygophyllaceae/Balanitaceae
Genus: *Balanites*
Species: *aegyptiaca*

1.5.2 Binomial name

Balanites aegyptiaca (L.) Del.

1.5.3 Common names

In English it is called Desert date; Jericho balsam; Egyptian balsam; Balanos; Zachum oil plant; soapberry tree, in Arabic, it is known as *Lalob, Hidjihi,* and *Heglig,* while in Hindi it is called *Hingota*.

1.5.4 Habitat

It is indigenous to all drylands south of Sahara and extending southwards (Hall and Walker 1991; Sidiyene 1996; Sands 2001). It is found in tropical and northern Africa; Syria; West Asia; Sudan; Egypt; neighboring parts of East and West Africa, particularly Senegal and Nigeria; Arabia; and Burma. In India, it is distributed in the drier parts of western Rajasthan and from Southeast Punjab to West Bengal and Sikkim (Amalraj and Shankarnarayan 1998; Anonymous 2001). It can be found in many kinds of habitat, tolerating a wide variety of soil types, from sand to heavy clay, and climatic moisture levels, from arid to subhumid. It is relatively tolerant of flooding, livestock activity, and wildfire (Von Maydell 1984).

1.5.5 Morphological Description

It is a multibranched, spiny shrub or tree up to 10 m tall. The crown is spherical, in one or several distinct masses. The trunk is short and often branching from near the base. Branches are pubescent, with axillary spreading spines up to 4 cm. The bark is dark brown to grey, deeply fissured. Branches are armed with stout yellow or green thorns up to 8 cm long. Leaves grow with two separate leaflets; leaflets are obovate, asymmetric, 2.5–6 cm long, bright green, and leathery, with fine hair when young. Flowers grow

Fig. 1.1 Balanites aegyptiaca a. *A semi-arid tree* b. *Green spiny branches* c. *Flowers* d. *Fruits*

in fascicles in the leaf axils and are fragrant and yellowish green. They are small, inconspicuous, and hermaphroditic and are pollinated by insects. The fruit is a rather long, narrow drupe, 2.5–7 cm long, and 1.5–4 cm in diameter. The young fruits are green and tomentose, which turn yellow and glabrous when mature. The pulp is bittersweet and edible. The seed, a pyrene stone, is 1.5–3 cm long, light brown, fibrous, and extremely hard. It makes up 50–60% of the fruit. There are 500–1,500 dry, clean seeds per kilogram (Anonymous 2001; Fig. 1.1).

1.5.6 Active Constituents

Phytochemical investigations on *B. aegyptiaca* led to the isolation of several classes of secondary metabolites, many of which expressed biological activities, such as coumarins, flavonoids, and steroidal saponins (Sarker et al. 2000). Saponins (diosgenin) are the major compounds which have been found in the various parts of the plant extracts (Pettit et al. 1991; Farid et al. 2002; Lindin 2002; Staerk et al. 2006).

Two furostanol glycosides and 6-methyl-diosgenin were obtained from the fruits (Kamel

1998). The fruit pulp also contains five steroidal saponins, designated as balanitisin A, B, C, D, and E. The kernel contains balanitisin F and G. Chemically, balanitisin A has been identified as diosgenin-3-O-α-D-glucopyranosyl (1–3)-O-α-D-glucopyranosyl-(1–4)-εL-rhamnopyranoside. The diosgenin content of the fruits varies from 0.3 to 3.8% (Anonymous 2001). The kernel is very rich in both oil (46.0–55.0%) and protein (26.0–34.0%) (Mohamed et al. 2002). The defatted seeds are a source of diosgenin, because of the fruit wall which contains 1.5% diosgenin on a fresh weight basis.

The roots and bark of the plant also have several steroidal saponins, yamogenin glycosides, which were isolated (Pettit et al. 1991). The root wood contains balanitisin H and the stem wood balanitisin I. More recently, five new steroidal glycosides were isolated from the roots of the plant (Farid et al. 2002). The diosgenin content of the roots varies from 0.3 to 1.5%. The leaves contain a saponin of diosgenin, stigmasterol, and a small amount of free diosgenin (Anonymous 2001).

1.5.7 Medicinal Properties and Uses

Various parts of the *B. aegyptiaca* have been used for folk medicines in many regions of Africa and Asia (Hall and Walker 1991; Neuwinger 1996; Mohamed et al. 2002). Literature has revealed antifeedant, antidiabetic, molluscicide, antihelmintic, anti-inflammatory, antimicrobial, analgesic, and contraceptive activities in various plant extracts (Liu and Nakanshi 1982; Ibrahim 1992; Rao et al. 1997; Gaur et al. 2009).

The bark, unripe fruits, and leaves of this plant are reported to have antifertility, purgative, and antidysentric properties. The kernel oil possesses antibacterial and antifungal properties and is reported to be used in the treatment of skin diseases, burns, excoriations, and freckles. The fruit and seed are considered useful in whooping cough and colic. The powdered seed is used for easy childbirth and the decoction of root is reported to be emetic (Anonymous 2001).

Recently, the study has endeavored to utilize esters of *B. aegyptiaca* as a fuel for diesel engine. Ester developed from balanites oil by the transesterification process, which has been investigated for its properties and engine performance (Deshmukh and Bhuyar 2009). Also, it has been used as a model for the utilization of bioresources in the Israeli Arava desert and potentially other similar areas for cost-effective biodiesel production (Chapagain et al. 2009).

1.5.8 Other Uses

Many parts of the plant are used as famine foods in Africa; the leaves are eaten raw or cooked, the oily seed is boiled to make it less bitter and eaten mixed with sorghum, and the flowers can be eaten. The tree is considered valuable in arid regions because it produces fruit even in dry times. The fruit can be fermented for the production of alcoholic beverages. The oil is used as cooking oil. The seed cake remaining after the oil is extracted is commonly used as animal fodder in Africa. The tree is managed through agroforestry. It is planted along irrigation canals and is used to attract insects for trapping. The wood is used to make furniture and durable items such as tools, and it is a low-smoke firewood and good charcoal. The smaller trees and branches are used as living or cut fences because they are resilient and thorny. The tree fixes nitrogen. The bark yields fibers, the natural gums from the branches are used as glue, and the seeds have been used to make jewelry and beads.

1.5.9 Conventional Propagation Methods and its Limitations

Conventionally, *Balanites* is propagated by seeds and through root suckers. These methods are not very successful being season dependent, space requiring, and cumbersome in nature. Another bottleneck is that the propagation through seeds or root sucker is also not very efficient in producing sufficient number of planting stocks due to poor germination of seeds, and root sucker pro-

Table 1.1 Plant tissue culture of *Balanites aegyptiaca* (L.) Del.

Mode of regeneration	Explant used	Response	References
Direct	Root	Adventitious shoots, plantlet regeneration	Gour et al. (2005); Varshney and Anis (2013)
Indirect	Cotyledon	Callus, multiple shoot formation, plantlet regeneration	Gour et al. (2007a, b)
Direct	Node	Axillary multiple shoot formation, plantlet regeneration	Ndoye et al. (2003); Siddique and Anis (2009a); Anis et al. (2010)

duction is strictly age dependent (Siddique and Anis 2009a). Therefore, we need to sort alternate methods like micropropagation for rapid multiplication of planting stock material.

- Tissue culture studies in *B. aegyptiaca* (L.) Del.

Although the plant has huge manifold applications for mankind, there are only a couple of reports available on micropropagation for its conservation and replenishment as shown in Table 1.1.

After reviewing the existing status of *B. aegyptiaca* (L.) Del. and the published protocols on its in vitro propagation where a lesser number of shoots has been obtained, it represented the need for more formal research effort on the subject. The book deals with the development of the improved protocols for mass propagation of *B. aegyptiaca* (L.) Del. with the following objectives: (1) to establish and proliferate axenic cultures from both juvenile and mature explants, (2) to formulate culture conditions for regeneration and multiplication and regenerant differentiation in somatic tissues, (3) to select the best-suited media for direct regeneration, (4) to understand the anatomical processes of shoot–bud differentiation under the influence of plant growth regulators, (5) to successfully acclimatize the regenerants using in vitro- and ex vitro-hardening methods, (6) to optimize the technique of synthetic seed production and their conversion into plantlets, (7) to study the effects of different days of acclimatization on the physiological and biochemical parameters of regenerated plantlets, and (8) to check the clonal fidelity of micropropagated plantlets using the inter-simple sequence repeat (ISSR) analysis.

References

Ahuja MR (1993) Micropropagation a la carte. In: Ahuja MR (ed) Micropropagation of woody plants, forestry series, vol 41. Kluwer Academic, Dordrecht, Netherlands, pp 3–9

Amalraj VA, Shankanarayan KA (1998) Ecological distribution of *Balanites roxburghii* plant in arid Rajasthan. J Trop Fores 2:183–187

Anis M, Ahmad N, Siddique I, Varshney A, Naz R, Perveen S, Khan Md I, Ahmed Md R, Husain MK, Khan PR, Aref IM (2012) Biotechnological approaches for the conservation of some forest tree species. In: Jenkins AJ (ed) Forest decline: causes and impacts. Nova Publishers, Inc., pp 1–39

Anis M, Varshney A, Siddique I (2010) In vitro clonal propagation of *Balanites aegyptiaca* (L.) Del. Agrofores Syst 78:151–158

Anonymous (2001) The wealth of India: a dictionary of Indian raw materials and industrial products, Publication and Information Directorate, CSIR, New Delhi. 2:3–5

Chapagain B, Yehoshua Y, Wiesman Z (2009) Desert date (*Balanites aegyptiaca*) as an arid lands sustainable bioresource for biodiesel. Bio Technol 100:1221–1226

Deshmukh SJ, Bhuyar LB (2009) Transesterified Hingan (Balanites) oil as a fuel for compression ignition engines. Bio Bioen 3:108–112

Evans J (1998) The sustainability of wood production in plantation forestry. Unasylva 49:47–52

Farid H, Haslinger E, Kumert O (2002) New steroidal glycosides from *Balanites aegyptiaca*. Helv Chim Acta 85:1019–1026

FRA (2001) Global Forest Resources Assessment (2000, FAO, Rome) www.fao.org/forestry/ fo/fra/index.jsp

Gaur K, Nema RK, Kori ML, Sharma CS, Singh V (2009) Anti-inflammatory and analgesic activity of *Balanites aegyptiaca* in experimental animals models. Int J Green Pharma pp 214–217

References

Giri CC, Shyankumar B, Anjaneyulu C (2004) Progress in tissue culture, genetic transformation and applications of biotechnology to trees: an overview. Trees 18:115–135

Gour VS, Emmanuel CJSK, Kant T (2005) Direct *in vitro* shoot morphogenesis in desert date- *Balanites aegyptiaca* (L.) Del. from root segments. Multipurpose trees in the tropics: management and improvement Strategies, pp 701–704

Gour VS, Sharma SK, Emmanuel CJSK, Kant T (2007a) A rapid *in vitro* morphogenesis and acclimatization protocol for *Balanites aegyptiaca* (L.) Del.- a medicinally important xerophytic tree. J Plant Biochem Biotechnol 16:151–153

Gour VS, Sharma SK, Emmanuel CJSK, Kant T (2007b) Stomata and cholorophyll content as marker traits for hardening of *in vitro* raised *Balanites aegyptiaca* (L.) Del. plantlets. Natl Acad Sci Lett India 30:45–47

Gurumurthi K (1994) Forest tree improvement in India- Baseline strategy, in UNDP/FAO project on improved productivity of man-made projects through application of technological advance in tree breeding and propagation (FAO, Rome), pp 1–41

Hall JB, Walker DH (1991) *Balanites aegyptiaca*; A monograph. School of Agricultural and Forest Science, University of Wales, Banger, pp 1–12

Harry IS, Thorpe TA (1994) *In vitro* culture of forest trees. In: Vasil IK, Thorpe TA (eds) Plant cell and tissue culture. Kluwer Academic Publishers, Dordrecht, 539–560

Hendrix SD, Kyhl JF (2000) Population size and reproduction in *Phlox pilosa*. Conserv Biol 14:304–313

Ibrahim AM (1992) Anthelmintic activity of some Sudanese medicinal plants. Phytother Res 6:155–157

IUCN (2010) IUCN Red List of Endangered Species In: http://www.iucnredlist.org/

IUCN (ed) (1998) The 1997 IUCN Red List of Threatened Plants. Gland, Switzerland and Cambridge, UK; Compiled by the World Conservation Monitoring Centre, international Union for the Conservation of Nature (IUCN) 862

Kamboj VP (2000) Herbal medicines. Curr Sci 78:35–39

Kamel MS (1998) A furostanol saponin from fruits of *Balanites aegyptiaca*. Phytochem 48:755–757

Kanonshi PJ (1997) Plantation forestry for the 21st century, XIth World for Congress, held on 13–22 Oct, 1997 (Antalya, Turkey), pp 23–34

Krishnan PN, Decruse SW, Radha RK (2011) Conservation of medicinal plants of Western Ghats, India and its sustainable utilization through *in vitro* technology. Vitro Cell Dev Biol-Plant 47:110–122

Lakshmisita G, Raghavaswamy BV (1998) Application of biotechnology in forest trees clonal multiplication of sandal wood, rose wood, teak, eucalypts and bamboos by tissue culture in India. In: Puri (ed) Tree improvement. Oxford, New Delhi, pp 233–248

Lindin T (2002) Isolation and structure determination of saponins from *Balanites aegyptiaca*. M.Sc. Thesis. The Royal Danish School of Pharmacy, Denmark

Liu HW, Nakanishi K (1982) The structure of Balanites; potent molluscides isolated from *Balanites aegyptiaca*. Tetrahedron 38:513–519

McCown BH (2000) Recalcitrance of woody and herbaceous perennial plants: dealing with genetic predetermination. Vitro Cell Dev Biol-Plant 36:149–154

Merritt DJ, Dixon KW (2003) Seed storage characteristics and dormancy of Australian indigenous plant species. In: Smith RD, Dickie JB, Linington SH, Pritchard HW, Probert RJ (eds) Seed conservation: turning science into practice. Royal Botanic Gardens Kew, Cromwell, London, pp 809–823

Merritt DJ, Turner SR, Clarke S, Dixon KW (2007) Seed dormancy and germination stimulation syndromes for Australian temperate species. Aust J Bot 55:336–344

Mohamed AM, Wolf W, Speiss W (2002) Physical, morphological and chemical characteristics, oil recovery and fatty acid composition of *Balanites aegyptiaca* Kernels. Plant Foods Hum Nutr 57:179–189

Ndoye M, Diallo I, Gassamaldia YK (2003) *In vitro* multiplication of the semi-arid forest tree, *Balanites aegyptiaca* (L.) Del. Afr J Biotechnol 2:421–424

Neuwinger HD (1996) African ethnobotany: poisons and drugs. Chapman and Hall, London, pp 884–890

Pandey D, Ball J (1998) The role of industrial plantations in future global fibre supplies. Unasylva 49:37–43

Pettit GR, Doubek DL, Herald DL, Numata A, Takahasi C, Fujiki R, Miyamoto T (1991) Isolation and structure of cytostatic steroidal saponins from the African medicinal plant *Balanites aegyptiaca*. J Nat Prod 54:1491–1502

Pierik RLM (1997) *In vitro* culture of higher plants. Martinus Nijhoff Publishers, Dordrecht

Rao MV, Shah KD, Rajani M (1997) Contraceptive efficacy of *Balanites roxburghii* pericarp extract in male mice (*Mus musculus*). Phytother Res 11:469–471

Sands MJ (2001) The desert date and its relatives: a revision of the genus Balanites. Kew Bulletin 56:1–128

Sarker SD, Bartholomew B, Nash RJ (2000) Alkaloids from *Balanites aegyptiaca*. Fitoterapia 71:328–330

Sharma M (2002) Agricultural biotechnology, in country case studies, edited by GJ Persley and LR MacIntyre (CAB International) 51–60

Siddique I, Anis M (2009a) Direct plant regeneration from nodal explants of *Balanites aegyptiaca* L. (Del.): a valuable medicinal tree. New Fores 37:53–62

Sidiyene EA (1996) Trees and shrubs in the Adrar Iforas, Mali. Editions de I'orslom, Paris, France.

Srivastava J, Lambert J, Vietmayer N (1995) Medicinal plants: a growing role in development agriculture and natural resources. Department of agriculture and forestry system. The World Bank, Washington, USA

Staerk D, Chapagain BP, Lindin T, Wiesman Z, Jaroszewski JW (2006) Structural analysis of complexsaponins of *Balanites aegyptiaca* by 800 MHz 1H NMR spectroscopy. Magn Reson Chem 44:923–998

Thorpe TA, Harry IS (1991) Application of micropropagation to forestry. In: Debergh PC, Zimmerman RH (eds) Micropropagation: technology and application. Kluwer Academic Publishers, Dordrecht, pp 311–336

Varshney A, Anis M (2013) Direct plantlet regeneration from segments of root of *Balanites aegyptiaca* Del. (L.)- a biofuel arid tree. Int J Pharm Bio Sci 4:987–999

Vines G (2004) Herbal harvests with a future: towards a sustainable source for medicinal plants, Plant Life International. www.plantlife.org.uk.

Von Maydell HJ (1984) Arbes et arbustes du Sahel: leurs caracteristiques et leurs utilisations. GTZ, Eschborn, 531

Williams CG, Savolainen O (1996) Inbreeding depression in conifers: implications for using selfing as a breeding strategy. Fores Sci 42:102–117

Yasodha R, Sumathi R, Gurumurthi K (2004) Micropropagation for quality propagule production in plantation forestry. Indian J Biotechnol 3:159–170

Review of Literature

2

Abstract

Gottlieb Haberlandt (Math Naturwiss 111:69–92), a German plant physiologist, for the first time initiated the work on tissue culture. His work arose as a research tool and attempts were made to culture the isolated, fully differentiated cells in nutrient medium in vitro as early as 1898. The theoretical basis of tissue culture lies in the cell theory given by Schleiden and Schwann (1838–1839). Practically, this technique stands on the concept of "totipotency," i.e., each cell has the ability to regenerate into a new plant. The field finds a wide range of applications starting from mass clonal propagation to plant improvement, molecular biology, bio-processing as well as a basic research tool. It has advanced the production in forestry and agriculture to many folds. There are a number of reviews published on tissue culture of woody and tree species which provide the wide-ranging micropropagation reports of various plant species. In this way, this chapter highlights the current review in tree tissue culture in vitro and micropropagation of other valuable plants, their significance, and the wide scope existing for investigations on mass multiplication and conservation of these plants.

2.1 Introduction

Plant biotechnology is founded on the demonstrated totipotency of plant cells, combined with the delivery, stable integration, and expression of transformed plants, and the Mendelian transmission of transgenes to the progeny. The concept of totipotency itself is inherent in the cell theory of Schleiden (1838) and Schwann (1839), which forms the basis of modern biology by recognizing the cell as the primary unit of all living organisms. The cell theory received much impetus from the famous aphorism of Virchow (1858), "Omnis cellula a cellula" (all cells arise from cells), and by the very persistent observation of Vochting (1878) that the whole plant body can be built up from ever so small fragments of plant organs. However, no sustained attempts were made to test the validity of these observations, until the beginning of the twentieth century because the required technologies did not exist and the nutritional requirements of cultured cells were not fully understood (Gautheret 1985).

Haberlandt (1902) was the first to conduct experiments designed to demonstrate the totipotency of plant cells by culturing chloroplast-containing differentiated cells from leaves of *Lamium purpureum*, cells from the petioles of *Eichhornia crassipes*, glandular hairs of *Pulmonaria* and *Urtica*, and stamen hairs of *Tradescantia* in

Knop's (1865) salt solution in hanging drop cultures. The cells grew in size, but did not undergo any cell divisions. He failed largely because of his unfortunate choice of experimental material and because of the inadequacy of nutrition provided by Knop's salt solution.

Even now, more than a century later, the culture of isolated leaf cells, and other materials used by Haberlandt, is either extremely difficult or impossible in most species. Nevertheless, based on his experiences, Haberlandt made some bold predictions. He advocated the use of embryo sac fluids (coenocytic liquid endosperm, such as coconut milk that was later widely and successfully used in tissue culture studies) for inducing cell divisions in vegetative cells and pointed to the possibility of successfully cultivating artificial embryos (i.e., somatic embryos, which are now the predominant means of plant regeneration in a wide range of species (Thorpe 1995; Vasil 1999)) from vegetative cells in nutrient solutions. It is for these reasons that Haberlandt is rightfully credited with being the founder of the science of plant cell culture.

Went's discovery of auxin (Went 1928) and its subsequent identification as indole-3-acetic acid (IAA) in 1934 (Went and Thimann 1937; Haagen-Smit 1951) opened the road to early successes in tissue culture (Gautheret 1983, 1985). Lyophilized leaf extract from leaves of dodder host plants (a preparation which probably contained auxin) and pure IAA were incorporated in culture media with mixed results starting about a decade after the discovery of this hormone (Fiedler 1936; Geiger-Huber and Burlet 1936; Gautheret 1935, 1937; Loo 1945a).

Skoog and Tsui (1951) reported the continued cell division and bud formation in the cultured pith tissues of tobacco on nutrient media containing adenine and high levels of phosphate. However, Jablonski and Skoog (1954) observed the occurrence of cell divisions when the explants containing vascular tissue were cultured only on nutrient medium. A variety of plant extracts, including coconut milk, were added to the nutrient medium in an attempt to replace vascular tissues and to identify the factors responsible for their beneficial effect. Among these, yeast extract was found to be the most effective and its active component was shown to have purine-like properties. This finding led to the addition of DNA to the medium which greatly enhanced cell division activity (Vasil 1959). These investigations resulted in the isolation of kinetin (6-furfurylaminopurine, Kn) from old samples of herring sperm DNA (Miller et al. 1955) and the understanding of the hormonal (auxin–cytokinin) regulation of shoot morphogenesis in plants (Skoog and Miller 1957).

Later experiments led to the isolation of naturally occurring as well as many synthetic cytokinins, the elucidation of their role in cell division and bud development, and their extensive use in the micropropagation industry related to their suppression of apical dominance resulting in the development of many axillary shoots.

The development of improved nutrient solutions, informed choice of plant material, and appreciation of the importance of aseptic cultures led to long-term or indefinite cultures of excised tomato roots and cambial tissues of tobacco and carrot, by White (1934a and b, 1939) in the USA and Gautheret (1934, 1939) and Nobecourt (1939) in France.

White (1943) and others believed that the nutrient solutions based on Knop's (1865) and other formulations neither provided optimal growth nor were stable or satisfactory over a wide range of pH values. These concerns led to the development of White's (1943) medium, which was widely used until the mid-1960s. During this period, a systematic study of mineral and other requirements of plant tissues grown in culture was carried out (Hildebrandt et al. 1946; Heller 1953), demonstrating the need for a greatly increased level of mineral salts in the medium (Ozias-Akins and Vasil 1985). In a similar study, designed to optimize the growth of cultured tobacco pith tissue, a marked increase in growth obtained by the addition of aqueous extracts or ash of tobacco leaves to White's medium was found to be caused largely by the inorganic constituents of the extracts, leading to the development of the first chemically defined and most widely used nutrient solution for plant tissue cultures (Murashige and Skoog 1962). The principal novel features of the Murashige and Skoog (MS) medium were the very high levels of inorganic constituents, chelated iron in order to make it more stable and available during the life of cultures, and a mixture of four vitamins and myo-inositol.

The discovery of the protoplast culture by Cocking in 1960, haploid plants from anther culture (Guha and Maheshwari 1966), mericloning (Morel 1964), and stages of tissue culture by Murashige (1974) further added aspects of the plant cell culture.

Though hardwood trees were among the first plants cultured in vitro (Gautheret 1940), the first complete plant from tissue culture of free living angiosperm tree species was reported several years later by Winton (1968) from leaf explants of black cotton wood (*Populus trichocarpa*) and by Wolter (1968) for *Populus tremuloides*. The tissue culture of perennial and woody species, being difficult to yield quick results because of their inherent slow-growing nature besides intractable regeneration potential, in addition to some other factors, naturally prompted less effort. But of late, the accent has shifted to a good extent to regenerate trees which used to pose insurmountable challenges in conventional practices of propagation.

Previous work with other woody species, including members of the *Pinus* and *Populus* genera, has provided robust knowledge about the need for a biotechnology-based breeding program in order to develop domesticated, and preferably clonal, elite varieties able to sustain the consistent and homogeneous production of high-quality plant material (Park et al. 1998; Giri et al. 2004; Boerjan 2005). Clonal forestry has enabled rapid genetic gain, thus opening up the possibility for domestication and massive proliferation of elite individuals within a relatively short time (Campbell et al. 2003; Merkle and Nairn 2005). Micropropagation has been used in breeding programs of the most important tropical forest genera, such as *Eucalyptus* (Pinto et al. 2002; Ruiz et al. 2005), *Tectona* (Hausen and Pal 2003), *Cedrela* (Nunes et al. 2002; Pena-Ramirez et al. 2010), *Acacia* (Vengadesan et al. 2002), *Gmelina* (Naik et al. 2003), *Azadirachta* (Quraishi et al. 2004; Morimoto et al. 2006), *Buchanania* (Sharma et al. 2005), *Simmondsia* (Bashir et al. 2007), *Pterocarpus* (Husain et al. 2010), and *Tabebuia donnell-smithii rose* (Gonzalez-Rodriguez et al. 2010).

2.2 Micropropagation

Schaeffer (1990) defined micropropagation as the in vitro clonal propagation of plants from shoot tips or nodal explants, usually with an accelerated proliferation of shoots during subcultures. Micropropagation is usually described as having the following four distinct stages (a stage "0" is added by some authors):

1. Stage "0": prepreparation of in situ donor material (fungicide and/or plant growth regulator (PGR) treatments, hedging, etiolation, etc.)
2. Stage "I": initiation (including surface sterilization) of explants
3. Stage "II": shoot multiplication (optimization of proliferation media)
4. Stage "III": root induction on microcuttings (in vitro or ex vitro)
5. Stage "IV": acclimatization of rooted shoots (or unrooted microcuttings, to ex vitro conditions)

Shoot culture remains the most widely used form of clonal tissue culture and with care can provide a ready source of disease-free and contaminant-free material (George 1993). Induction of shoots into culture (provided the initiation of viable sterile cultures is possible) involves manipulation of the growing medium components, chiefly PGRs to achieve optimal shoot multiplication (stage II) and root induction (stage III) prior to acclimatizing and transferring rooted shoots into the extra vitrum environment (stage IV).

Various degrees of difficulty can be encountered in stages I–IV between species and even between genotypes within a species (Naik et al. 2003). Inability to achieve sterile explants, poor explant performance due to oxidation, phenolic leakage, and the premature death of explants can be encountered in stage I (Lynch 1999). Lack of response to cytokinins, slow growth, abnormal growth, e.g., hyperhydric transformation, shoot miniaturization, or stunting, prolonged phenolic exudation, shoot necrosis, or excessive callusing may impede optimization of shoot multiplication media in stage II, especially with woody species (Benson 2000). Lack of response (poor or no root induction) to auxin(s), excessive callusing, or deterioration in overall shoot quality can be encountered in stage III, especially with woody plants (Lynch 1999).

In stage IV, transfer to soil, plants need physiological adjustment to the ex vitro environment. This entails physiological adjustment to reduced mineral nutrient loading, more variable temperatures, higher lighting levels, reduced humidity, reinstatement of waxy leaf coatings to prevent desiccation and resumption of stomatal function, and regaining root function to allow mass flow transpiration (Preece and Sutter 1991). In addition, plants must successfully survive the transition from primary dependence on medium sugar as the carbon source (photoheterotrophic or photomixotrophic) and become photoautotrophic again (Pospisilova et al. 1999a and b). All these changes need to occur relatively quickly for the plant to regain physiological competence and avoid a prolonged transition period in stage IV with subsequent risk of senescence resulting from oxidative trauma (Batkova et al. 2008) and pathogen infection (Williamson et al. 1998).

2.3 Various Approaches for Micropropagation

The methods that are available for propagation of plants in vitro are described in the following sections of the chapter. They are essentially as follows:
- By the multiplication of shoots from axillary buds

By the formation of adventitious shoots, and/or adventitious somatic embryos, either (a) directly on pieces of tissue organs (explants) removed from the mother plant or (b) indirectly from unorganized cells (in suspension cultures) or tissues (in callus cultures) established by the proliferation of cells within explants, or on semiorganized callus tissues or propagation bodies (such as protocorms or pseudobulbils) that can be obtained from explants (particularly those from certain specialized whole plant organs)

2.3.1 The Propagation of Plants from Axillary Buds or Shoots

The production of plants from axillary buds or shoots has proved to be the most generally applicable and reliable method of true-to-type in vitro propagation. The two methods commonly used are as follows:
- Shoot culture
- Single, or multiple, node culture

Both depend on stimulating precocious axillary shoot growth by overcoming the dominance of shoot apical meristems.

Robbins (1922) seems to have been the first person to have successfully cultured excised shoot tips on a medium containing sugar, but a significant shoot growth from vegetative shoot tip explants was first achieved by Loo (1945a, b, 1946a, b) in *Asparagus* and dodder plants. He has made several significant observations in *Asparagus* cultures showing that:
- Growth depended on sucrose concentration, higher levels being necessary in the dark than in the light.
- Explants, instead of being supported, could apparently be continued indefinitely (35 transfers were made over 22 months).
- Shoot tip culture afforded a way to propagate plant material (clones were established from several excised shoot apices).

This work failed to progress further because no roots were formed on the *Asparagus* shoots in culture. Honors for establishing the principles of modern shoot culture must therefore be shared between Loo and Ball.

Ball (1946) was the first person to produce rooted shoots from cultured shoot apices. His explants consisted of an apical meristem and two to three leaf primordia. There was no shoot multiplication but plantlets of *Tropaeolum majus* and *Lupinus alba* were transferred to soil and grown successfully.

Haramaki (1971) described the rapid multiplication of *Gloxinia* by shoot culture, and, by 1972, several reports of successful micropropagation by this method had appeared (Adams 1972; Haramaki and Murashige 1972). Since then, the number of papers on shoot culture published annually has increased dramatically and the method has been utilized increasingly for commercial plant propagation (George and Debergh 2008).

Currently, node culture is of value for propagating species that produce elongated shoots in culture, especially if stimulation of lateral bud break is difficult to bring about with available cytokinins. Nowadays, the technique becomes more and more popular in commercial micropropagation. The main reason is that it gives more guarantee for clonal stability. Indeed, although the rate of multiplication is generally less than that which can be brought about through shoot culture, there is less likelihood of associated callus development and the formation of adventitious shoots, so that stage II subculture carries very little risk of induced genetic irregularity (George and Debergh 2008). For this reason, node culture has been increasingly recommended by research workers as the micropropagation method least likely to induce somaclonal variation (Prakash and Van Staden 2008; Ahmad and Anis 2011; Asthana et al. 2011).

2.3.1.1 Multiple Shoots from Seeds

During the early 1980s, it was discovered that it was possible to initiate multiple shoot clusters directly from seeds. Seeds are sterilized and then placed onto a basal medium containing a cytokinin. As germination occurs, clusters of axillary and/or adventitious shoots ("multiple shoots") grow out and may be split up and serially subcultured on the same medium. High rates of shoot multiplication are possible. For instance, Hisajima (1982) estimated that ten million shoots of almond could be derived theoretically from one seed per year. Similarly, Malik and Saxena (1992a, b) found that 5–20-fold more shoots were induced when intact seeds of pea, chickpea, and lentil were exposed to a medium with high concentrations of cytokinins such as benzyladenine (BA) or thidiazuron (N-fenil-N′-1,2,3-thidiazol-5-il-urea, TDZ), compared to isolated explants such as cotyledonary nodes. Malik and Saxena (1992b) hypothesized that the high regeneration potential of seeds is caused by the physiological structural integrity of the explants.

It is likely that multiple shoots can be initiated from the seeds of many species, particularly dicotyledons. The technique is effective in both herbaceous and woody species: soybean (Hisajima and Church 1982), sugar beet (Powling and Hussey 1981), almond (Hisajima 1981), walnut (Rodriguez 1982), chickpea, lentil (Malik and Saxena 1992a), *Pisum sativum* (Malik and Saxena 1992b; Zhihui et al. 2009), *Murraya koenigii* (Bhuyan et al. 1997), *Litchi chinensis* (Das et al. 1999), and *Sterculia urens* (Hussain et al. 2008).

2.3.2 Propagation by Adventitious Shoot Organogenesis

During adventitious organogenesis, new organs (shoots, roots) can develop on explants of different plant tissues such as leaves, stems, and roots. Leaves and hypocotyls, for instance, do not have any apparent preexisting meristems, and therefore most plant cells could be considered totipotent.

The huge amount of variability seen in the frequency of organogenesis between varieties and species suggests that it is the proportion of cells that are receptive to in vitro culture conditions that vary. Organogenesis in vitro consists of several factors, such as PGR perception and transduction, redifferentiation after dedifferentiation of differentiated cells, organization for specific organ primordial and meristems, etc. The process depends on external and internal factors, such as exogenously applied PGRs, and the ability of plant tissue to perceive these PGRs.

According to the requirement for a specific PGR balance, three phases of organogenesis are distinguishable (Sugiyama 1999):
1. Competence (dedifferentiation): Cells become able to respond to hormonal signals of organ induction (Howell et al. 2003). Dedifferentiation means the acquisition of organogenic competence. Wounding usually triggers dedifferentiation in tissue culture (Sugiyama 1999).
2. Determination: The competent cells in explants are determined for specific organ formation and this process is influenced by a specific PGR balance. From competent cells,

adventitious shoot formation could be induced by cytokinins (Gahan and George 2008).

3. Morphogenesis: This proceeds independently of exogenously added PGRs (Sugiyama 1999). However, Yancheva et al. (2003) found that the type of auxin and the length and timing of its application are critical for both activation and progression of the plant cell developmental program.

There are two distinct pathways of adventitious meristem formation: direct and indirect. During the direct pathway, the formation of a meristem proceeds without intermediate proliferation of undifferentiated callus tissue. However, meristems can be formed indirectly from unspecialized and dedifferentiated cells of callus or suspension culture (Yancheva et al. 2003; Gahan and George 2008).

In certain species, adventitious shoots which arise directly from the tissues of the explants (and not within previously formed callus) can provide a reliable method for micropropagation. The main advantages of micropropagation by direct adventitious shoot regeneration are as follows:

- Initiation of stage I cultures and stage II shoot multiplication are more easily achieved than by shoot culture.
- Rates of propagation can be high, particularly if numerous small shoots arise rapidly from each explant.

However, the induction of direct shoot regeneration depends on the nature of the plant organ from which the explants were derived and is highly dependent on plant genotype.

Regeneration via adventitious shoots in woody species has been reported in *Dalbergia latifolia* (Lakshmisita et al. 1986), *Aegle marmelos* (Islam et al. 1993), *Sesbania grandiflora* (Detrez et al. 1994), *Liquidambar styraciflua* (Kim et al. 1997), *Prunus avium* (Hammatt and Grant 1998), *D. sissoo* (Pradhan et al. 1998; Pattnaik et al. 2000; Chand et al. 2002), *Azadirachta indica* (Eswara et al. 1998; Sharma et al. 2002), *Salix nigra* (Lyyra et al. 2006), *Paulownia tomentosa* (Corredoira et al. 2008), *Platanus occidentalis* (Sun et al. 2009), *Jatropha curcas* (Kumar et al. 2010), *Cassia angustifolia* (Siddique et al. 2010),*Tabebuia donnell-smithii rose* (Gonzalez-Rodriguez et al. 2010), *Pterocarpus marsupium* (Husain et al. 2010), and *Albizia lebbeck* (Perveen et al. 2011).

2.3.2.1 Regeneration from Roots

Among the possible initial explants, roots have proven to be highly regenerative explants for in vitro regeneration in different species, including forest ones (George 1993). According to Morton and Browse (1991), root explants are advantageous over other explants in terms of their easy manipulation, higher regeneration, and excellent susceptibility for *Agrobacterium* transformation. Roots have also received considerable attention as a potential production system for stable metabolite production (Zobayed and Saxena 2003). Besides being useful for micropropagation, the root culture could be successfully applied for germplasm preservation (Chaturvedi et al. 2004).

Since the pioneering work on the establishment of root culture of tomato (White 1934a), roots of many plants are capable of independent growth to form secondary roots in culture (Butcher and Street 1964) and regenerate shoots. These shoots arise de novo from pericycle, as demonstrated in *Convolvulus* (Torrey 1958). Initiation of shoots from root explants has been described in several plant species, indicating a possibility of developing regenerative excised root culture for mass multiplication and their germplasm preservation, viz., *Citrus mitis* (Sim et al. 1989), *Citrus aurantifolia* (Bhat et al. 1992), *Averrhoa carambola* (Kanthrajah et al. 1992), *Lonicera japonica* (Georges et al. 1993), *Acacia albida* (Ahee and Duhoux 1994), *Albizia julibrissin* (Sankhla et al. 1994), *Aeschynomene sensitiva* (Nef-Campa et al. 1996), *Azadirachta indica* (Sharma et al. 2002; Shahin-un-zamam et al. 2008; Arora et al. 2010), *Shorea robusta* (Chaturvedi et al. 2004), *Melia azedarach* (Vila et al. 2005), *Populus alba* (Tsvetkov et al. 2007), *Cleome rosea* (Simoes et al. 2009), *Swertia chirata* (Pant et al. 2010), *Passiflora edulis* (Viana da Silva et al. 2011), and *Albizia lebbeck* (Perveen et al. 2011).

2.4 Factors Affecting In Vitro Shoot Regeneration and Growth of Plants

Several factors, such as genotype, type of explants, PGRs, type of media, and in vitro conditions before and after the regeneration process, can influence the success of in vitro shoot regeneration.

2.4.1 Explant Type

The tissue which is obtained from the plant to culture is called an explant. Based on work with certain model systems, particularly tobacco, it has often been claimed that a totipotent explant can be grown from any part of the plant. In many species, explants of various organs vary in their rates of growth and regeneration, while some do not grow at all. Also, the risk of microbial contamination is increased with an inappropriate explant. Thus, it is very important that an appropriate choice of explant be made prior to tissue culture.

The specific differences in the regeneration potential of different organs and explants have various explanations. The significant factors include differences in the stage of the cells in the cell cycle, the availability of or ability to transport endogenous growth regulators, and the metabolic capabilities of the cells. The most commonly used tissue explants are the meristematic ends of the plants such as the stem tip, axillary bud tip, and root tip. These tissues have high rates of cell division and either concentrate or produce the required growth-regulating substances including auxins and cytokinins (Akin-Idowu et al. 2009).

Some studies have shown that explant characteristics such as type, source, genotype, and history affect the success and commercial viability of tissue culture systems (Bhau and Waklu 2001; Chan and Chang 2002; Hoy et al. 2003). The effect of explant type on successful tissue culture of various plants has been reported (Gubis et al. 2003; Blinstrubiene et al. 2004; Tsay et al. 2006; Gitonga et al. 2010).

Moreover, juvenile plants are an excellent explant source to achieve successful in vitro propagation of tropical forest trees. Nodal segments have been widely used for in vitro shoot proliferation of woody plants such as *Syzygium cuminii* (Jain and Babbar 2000), *Terminalia chebula* (Shyamkumar et al. 2003), *Oroxylum indicum* (Dalal and Rai 2004), *Holarrhena antidysenterica* (Kumar et al. 2005), *Boswellia ovalifoliolata* (Chandrasekhar et al. 2005), *Tectona grandis* (Shirin et al. 2005), *Pterocarpus santilinus* (Rajeswari and Paliwal 2006), miracle berry (Ogunsola and Ilori 2007), *Stereospermum personatum* (Shukla et al. 2009), *Vitex negundo* (Ahmad and Anis 2011), and *Sapindus trifoliatus* (Asthana et al. 2011). This is probably due to the readily available axillary buds in nodal segments that only require a trigger for bud break in contrast to root, leaf, and cotyledonary tissue that would otherwise require initiation of adventitious buds (Lombardi et al. 2007, Sivanesan et al. 2007) and somatic embryos (Naz et al. 2008, Husain et al. 2010) before any shoot regeneration is achieved.

The different explants such as nodal segments, cotyledonary nodes, hypocotyls, roots, and cotyledons selected also could influence the rate of shoot regeneration in many trees including different species of *Acacia*, teak, *Eucalyptus*, *Salix tetrasperma*, *V. negundo*, *P. marsupium*, and *Albizia lebbeck* (Vengadesan et al. 2002; Yasodha et al. 2004; Anis et al. 2012).

Different age of explants may have different levels of endogenous hormones and, therefore, the age of explants would have a critical impact on the regeneration success; these results have been reported in many plants, including *Prunus* (Mante et al. 1989), *Lachenalia* (Niederwieser and Van Staden 1990), *Cydonia oblonga* (Baker and Bhatia 1993), *Aegle marmelos* (Islam et al. 1993), *Malus* (Famiani et al. 1994), *Cercis canadensis* (Distabanjong and Geneve 1997), *Morus alba* (Thomas 2003), and *Sapindus trifoliatus* (Asthana et al. 2011).

The genotype, in addition to the type and age of the explants, is a key criterion in determining the material suitable for micropropagation (Kunze 1994; Tang and Guo 2001; Tereso et al. 2006; Cortizo et al. 2009).

2.4.2 Plant Growth Regulators

Some chemicals occurring naturally within plant tissues (i.e., endogenously) have a regulatory rather than a nutritional role in growth and development. These compounds, which are generally active at very low concentrations, are known as plant hormones (or plant growth substances). Synthetic chemicals with physiological activities similar to plant growth substances or compounds having an ability to modify plant growth by some other means are usually termed PGRs.

There are several recognized classes of plant growth substance. Until relatively recently, only five groups were recognized, namely:
- Auxins
- Cytokinins
- Gibberellins
- Ethylene
- Abscisic acid (ABA)

Auxins and cytokinins are by far the most important plant growth substances for regulating growth and morphogenesis in plant tissue and organ cultures; in these classes, synthetic regulators have been discovered with a biological activity, which equals or exceeds that of the equivalent natural growth substances.

No chemical alternatives to the natural gibberellins or ABA are available, but some natural gibberellins are extracted from cultured fungi and are available for use as exogenous regulants.

2.4.2.1 Auxins

Auxins are very widely used in plant tissue culture and usually form an integral part of the nutrient media. Auxins promote, mainly in combination with cytokinins, the growth of calli, cell suspensions, and organs and also regulate the direction of morphogenesis.

At the cellular level, auxins control basic processes such as cell division and cell elongation. Since they are capable of initiating cell division, they are involved in the formation of meristems giving rise to either unorganized tissue or defined organs. The choice of auxins and the concentration administered depend on:
- The type of growth and/or development required
- The rate of uptake and of transport of the applied auxin to the target tissue
- The inactivation (oxidation and/or conjugation) of auxin in the medium and within the explants
- The sensitivity of the plant tissue to auxin (and other hormones as well)
- The interaction, if any, between applied auxins and the natural endogenous substances

The most commonly detected natural auxin is IAA which may be used in plant tissue culture media, but it tends to be oxidized in culture media and is rapidly metabolized within plant tissues. However, for many purposes, it is necessary or desirable to use one of the many synthetic analogues of IAA. These analogues have different structures but similar biological properties and are also called auxins. The synthetic auxins 2,4-dichlorophenoxyacetic acid (2,4-D), α-naphthalene acetic acid (NAA), and indole-3-butyric acid (IBA) are commonly used in the tissue cultures.

Indeed, all active auxins are weak organic acids. The relative degree of activity of individual auxins in different growth processes is very variable. It differs not only from plant to plant but also from organ to organ, tissue to tissue, cell to cell, and also with the age and physiological state of the plant (tissue; Davies 2004).

In tissue culture, depending on other hormones present in the medium, changes in auxin concentrations may change the type of growth, e.g., stimulation of root formation may switch to callus induction. In this respect, each tissue culture system is unique, and the effects of different concentrations of auxins and other hormones must be tested for each case individually, and only to some extent, the results can be transferred to other cultures.

Plants—like other higher organisms—have to possess intra-organismal communication system(s) working over relatively long distances. As no nervous system is present, the main signaling systems are hormone dependent (Libbenga and Mennes 1995). Auxins are a component of such systems. Auxins and cytokinins impact at several levels in many different processes of plant development.

The ability of auxins (together with cytokinins) to manage key events in plant morphogenesis was documented, among others, by Skoog and Miller's (1957) discovery of the regulation of organogenesis in vitro by means of the auxin to cytokinin ratio in culture media. It has been further supported by recent investigations on the relationships between auxin and cytokinin levels and the morphogenetic response of various plants (Li et al. 1994; Leyser et al. 1996; Centeno et al. 1996).

A range of auxins in combination with cytokinins played a vital role in multiple shoot regeneration in many tree species (Vengadesan et al. 2002; Giri et al. 2004; Anis et al. 2012). Addition of low levels of auxins along with cytokinin is known to increase shoot numbers in many plant species like *Wrightia tinctoria* (Purohit and Kukda 1994), *Rauwolfia micrantha* (Sudha and Seeni 1996), *Sapium sebiferum* (Siril and Dhar 1997), *Gmelina arborea* (Tiwari et al. 1997), *Gymnema sylvestre* (Reddy et al. 1998), and *Tectona grandis* (Tiwari et al. 2002; Shirin et al. 2005). Moreover, the role of auxins in root development is well established and the effect of various auxins on rooting of the excised microshoots in vitro and ex vitro has been discussed in the section 2.5 "Rooting" of this chapter.

2.4.2.2 Cytokinins

Among PGRs, cytokinins have proven to be the most important factor affecting shoot regeneration, and their significant effects may be related to the histological changes in induced tissues (Magyar-Tabori et al. 2010).

Cytokinins are N^6-substituted adenines with growth regulatory activity in plants that promote cell division and may play a role in cell differentiation (McGaw and Burch 1995). Cytokinins added to the medium are very important during tissue culture of plants because they induce division and organogenesis (Howell et al. 2003) and affect other physiological and developmental processes (Heyl and Schmulling 2003; Ferreira and Kieber 2005; Van Staden et al. 2008).

The success of a culture is affected by the type and concentration of applied cytokinins, because their uptake, transport, and metabolism differ between varieties and they can interact with endogenous cytokinins of an explant (Werbrouck et al. 1996; Strnad et al. 1997; Van Staden et al. 2008).

There are two main classes of cytokinins according to the chemical structure of the side chain: isoprenoid and aromatic cytokinins, which differ in their biochemistry, receptors, biological activity, and their metabolism (Strnad et al. 1997; Werbrouck et al. 1996; Van Staden et al. 2008). Considering natural cytokinins, BA or sometimes Kn (Barciszewski et al. 1999) is most frequently used in tissue culture systems. Vengadesan et al. (2002) reported that BA singly was common in most of the in vitro micropropagation systems of different species of *Acacia* irrespective of the explants and media type. Similarly, the positive effect of BA on shoot multiplication of teak has been reported by a number of researchers (Gupta et al. 1980; Devi et al. 1994; Shirin et al. 2005).

Likewise, the superiority of BA over other cytokinins on multiple shoot bud differentiation has been demonstrated in a number of cases (Jeong et al. 2001; Loc et al. 2005, Phulwaria et al. 2012). Bhattacharya and Bhattacharya (1997) developed an in vitro culture protocol for *Jasminum officinale* using BA at an elevated level (17.76 µM) in MS medium. Similarly, Purkayastha et al. (2008) obtained maximum number of shoots at 10 µM BA. A considerable higher range of BA (22.2 µM) was used by Koroch et al. (1997) for the micropropagation of *Hedeoma multiflorum*. However, lower concentrations of BA (2.22 µM) were applied with full-strength MS medium by Elangomathvan et al. (2003) to obtain multiple shoots in *Orthosiphon spiralis*. Nayak et al. (2007) investigated the effect of BA at low concentrations on shoot proliferation in *Aegle marmelos* and found optimal response in MS medium supplemented with 6.6 µM BA. Gokhale and Bansal (2009) obtained multiple shoots directly from apical and axillary buds of *O. indicum* in MS medium amended with 4.43 µM BA.

Above and beyond the utilization of BA in tissue culture, Kn has also found to be very effective in establishing in vitro regenerative protocol for many plant species such as *Picrorhiza kurroa* (Lal et al. 1988), *Hemidesmus indicus* (Pattnaik and Debata 1996), *Limonium cavanillesii* (Amo-Marco and Ibanez 1998), *Eclipta alba,* and *Eupatorium adenophorum* (Borthakur et al. 2000). In addition, the application of this growth regulator for multiple shoot production has been successful in a number of medicinally important plant species including *Alpinia galanga* (Borthakur et al. 1999), *Eurycoma longifolia* (Hussein et al. 2005), *Tinospora cordifolia* (Gururaj et al. 2007), *Ricinus communis* (Chaudhary and Sood 2008), *Rauvolfia tetraphylla* (Harisaranraj et al. 2009), *Thymus vulgaris* and *T. longicaulis* (Ozudogru et al. 2011).

Moreover, 2-isopentenyladenine (2-iP) is reported to be the best cytokinin for shoot multiplication in blueberry by Cohen (1980) and in garlic by Bhojwani (1980). Chattopadhyay et al. (1995) achieved rapid micropropagation protocol for *Mucuna pruriens* using 2-iP. Mills et al. (1997) mentioned that 2-iP at a higher concentration of 30.5 mg dm^{-3} was optimum for differentiating maximum number of shoots in *Simmondsia chinensis*. However, Taha et al. (2001) reported that the shoot bud proliferation ability of date palm shoot tips was strongly enhanced by low concentration of 2-iP (3 mg dm^{-3}). Similarly, Jakola et al. (2001) obtained best results in *Vaccinium myrtillus* and *V. vitis-idaea* utilizing higher levels of 2-iP (49.2 and 24.6 µM) on modified MS medium, while lower concentrations (12.3 or 24.6 µM) were recommended by Pereira (2006) for other species of *Vaccinium cylindraceum* micropropagation on Zimmermann and Broome medium. In case of *Rhododendrons,* Vejsadova (2008) found the highest shoot multiplication rate on MS medium amended with 2-iP. Similarly, Singh and Gurung (2009) proved 2-iP to be the most effective cytokinin in comparison with BA or Kn for multiple shoot induction in *Rhododendron maddeni*.

There are also some important synthetic cytokinins, such as TDZ. The activity of TDZ varies widely depending on its concentration, exposure time, the cultured explant, and the species tested (Murthy et al. 1988). Like other synthetic cytokinins, it appears less susceptible to enzymatic degradation in vivo than other naturally occurring amino purine cytokinins and has been found to be effective at very low concentrations (0.0091–3.99 µM) for micropropagation of several species (Lu 1993). However, it has been used at higher concentrations (2.27–145.41 µM) for propagation of some species including *Zanthoxylum rhetsa* (Augustine and D'Souza 1997), a mature forest tree species. It has been shown to induce high bud regeneration rates in comparison to purine-based cytokinins and also has the capability of fulfilling both the cytokinin and auxin requirements of regeneration responses in a number of woody plants (Mok et al. 2005; Jones et al. 2007). However, it can also cause undesirable side effects, such as inhibited shoot elongation and rooting, fasciated shoots, and hyperhydricity.

In woody plants, TDZ has been shown to be suitable for micropropagation and regeneration of recalcitrant species or genotypes (Huetteman and Preece 1993; reviewed in Durkovic and Misalova 2008). A perusal of literature reveals that it has successfully been used to induce axillary or adventitious shoot proliferation in a number of plant species including herbaceous, perennials, and tree species such as *Cassia angustifolia* (Siddique and Anis 2007a and b; Parveen and Shahzad 2011), *P. marsupium* (Husain et al. 2007), *Pongamia pinnata* (Sujatha and Hazra 2007), *Cardiospermum halicacabum* (Jahan and Anis 2009), and *Bacopa monnieri* (Ceaser et al. 2010).

- Cytokinins and auxins in synergy

Since the action of cytokinin and auxin has been linked from early studies, they are known to interact in several physiological and developmental processes, including apical dominance, control of cell cycle, lateral root initiation, regulation of senescence, and vasculature development (Coenen and Lomax 1997; Swarup et al. 2002).

Exogenous applications of cytokinins and auxins have been known to be important for shoot induction and elongation of many plant species in vitro (George 1993). Various successful combinations have been reported, such as BA+IAA for *Aegle marmelos* (Islam et al. 1993), *Tecomella undulata* (Nandwani et al. 1995), *Paulownia* species (Rao et al. 1996; Bergmann and Moon 1997), *Tectona grandis* (Tiwari et al. 2002), *O. indicum* (Dalal and Rai 2004), *Malus zumi* (Xu et al. 2008), *Terminalia bellirica* (Phulwaria et al. 2012); BA + IBA for *G. arborea* (Sukartiningsih et al. 1999), *J. curcas* (Shrivastava and Banerjee 2008), *L. styraciflua* (Durkovic and Lux 2010); BA+NAA for *Syzygium alternifolium* (Sha Valli Khan et al. 1997), *Acacia catechu* (Hossain et al. 2001), *Zyziphus jujuba* (Hossain et al. 2003), *Tectona grandis* (Shirin et al. 2005), *Cornus mas* (Durkovic 2008),*Teucrium fruticans* (Frabetti et al. 2009), *Celastrus paniculatus* (Martin et al. 2006), and *Acacia auriculiformis* (Girijashankar 2011).

Also, elevated level of shoot multiplication and proliferation rate was achieved in *Myrica esculenta* using Kn in concurrence with NAA by Bhatt and Dhar (2004). Complete plantlets of *R. communis* were successfully raised by Chaudhary and Sood (2008) while applying Kn in combination with NAA in MS medium. Kn–NAA synergism and their triggering effect on shoot bud induction and multiplication have established rightly in *Cephaelis ipecacuanha* (Jha and Jha 1989), *Gossypium hirsutum* (Rauf et al. 2004), *Cordia verbenacea* (Lameira and Pinto 2006), *Alpinia officinarum* (Selvakkumar et al. 2007), and *Gardenia jasminoides* (Duhoky and Rasheed 2010). In addition, Iapichino and Airo (2008) reported that the addition of 2-iP with IAA is found to be the suitable PGR regime for propagation in *Metrosideros excelsa*.

2.4.2.3 Gibberellins

Plant tissue cultures can generally be induced to grow and differentiate without gibberellins, although gibberellic acid (GA_3) may become an essential ingredient of media for culturing cells at low densities (Stuart and Street 1971). GA_3 is known to break the dormancy of several types of seeds at a critical concentration. It stimulates seed germination via the synthesis of α-amylase and other hydrolases (Shepley et al. 1972). Thus, in recent papers, GA_3 has been used to break dormancy and morphogenesis (Chaturvedi et al. 2004; Shahzad et al. 2007; Parveen et al. 2010; Balaraju et al. 2011).

GA_3 is added to the medium, together with auxin and cytokinin, in stages I and II of shoot cultures of certain plants. At stage I, its presence can improve establishment; for example, De Fossard and De Fossard (1988) found that the addition of GA_3 to the medium was useful to initiate growth in cultures from adult parts of trees of the family Myrtaceae. Additions of GA_3 with BA caused high-frequency bud break and shoot multiplication in apical shoot buds and nodal explants of *Morus cathayana* (Pattnaik and Chand 1997). Moreover, the benefits of using GA_3 singly or in combination with other PGRs in the culture medium for shoot multiplication have been well documented in a number of plant species (Kotsias and Roussos 2001, Farhatullah and Abbas 2007, Moshkov et al. 2008).

2.4.3 Medium pH Levels

The relative acidity or alkalinity of a solution is assessed by its pH. This is a measure of the hydrogen ion concentration; the greater the concentration of H^+ ions (actually H_3O^+ ions), the more acidic the solution. As pH is defined as the negative logarithm of hydrogen ion concentration, acidic solutions have low pH values (0–7) and alkaline solutions have high values (7–14). To judge the effect of medium pH, it is essential to discriminate between the various sites where the pH might have an effect: (1) in the explants, (2) in the medium, and (3) at the interface between explants and medium (Thorpe et al. 2008).

According to Thorpe et al. (2008), the pH of a culture medium must be such that it does not disrupt the plant tissue. Within the acceptable limits, the pH also:

- Governs whether salts will remain in a soluble form
- Influences the uptake of medium ingredients and plant regulator additives
- Has an effect on chemical reactions (especially those catalyzed by enzymes)
- Affects the gelling efficiency of agar

This means that the effective range of pH for media is restricted. As will be explained, medium pH is altered during culture, but as a rule of thumb, the initial pH is set at 5.5–6.0. In culture media, detrimental effects of an adverse pH are generally related to ion availability and nutrient uptake rather than cell damage.

The pH of the medium has an effect on the availability of many minerals (Scholten and Pierik 1998). In general, the uptake of negatively charged ions (anions) is favored at acidic pH, while that of cations (positively charged) is best when the pH is increased. As mentioned earlier, the relative uptake of nutrient cations and anions will alter the pH of the medium. The release of hydroxyl ions from the plant in exchange for nitrate ions results in media becoming more alkaline; when ammonium ions are taken up in exchange for protons, the media become more acidic.

One of the chief advantages of having both NO_3^- and NH_4^+ ions in the medium is that uptake of one provides a better pH environment for the uptake of the other. The pH of the medium is thereby stabilized. Uptake of nitrate ions by plant cells leads to a drift towards an alkaline pH, while NH_4^+ uptake results in a more rapid shift towards acidity (George 1993). In media containing both NO_3^- and NH_4^+ with an initial pH of 5–6, the preferential uptake of NH_4^+ causes the pH to drop during the early growth of the culture. This results in increased NO_3^- utilization (Martin and Rose 1976). The final pH of the medium depends on the relative proportions of NO_3^- and NH_4^+ (Gamborg et al. 1968).

It has been reported that medium pH influences developmental processes in tissue culture, among other regenerative processes: xylogenesis in *Citrus* and *Zinnia elegans* (Khan et al. 1986; Roberts and Haigler 1994), androgenesis in winter triticale and wheat (Karsai et al. 1994), adventitious bud regeneration in tobacco (Pasqua et al. 2002), and adventitious root formation in apple (Harbage et al. 1998).

Changes in the pH of a medium do, however, vary from one kind of plant to another. In a random sample of papers on micropropagation, the average initial pH adopted for several different media was found to be 5.6 (mode 5.7) but adjustments to as low as 3.5 and as high as 7.1 had been made. Kartha (1981) found that pH 5.6–5.8 supported the growth of most meristem tips in culture and that cassava meristems did not grow for a prolonged period on a medium adjusted to pH 4.8. Shoot proliferation in *Camellia sasanqua* shoot cultures was best when the pH of a medium with MS salts was adjusted to 5–5.5. Many plant cells and tissues in vitro will tolerate pH in the range of about 4.0–7.2; those inoculated into media adjusted to pH 2.5–3.0 or 8.0 will probably die (Butenko et al. 1984). Bhatia and Ashwath (2005) reported that a high pH above 6.0 produces a very hard medium and a pH lower than 5.0 does not sufficiently solidify the medium.

Parliman et al. (1982) tested the effect of different medium pH values (3.5, 4.5, 5.5, and 6.5) in *Dionea muscipula* and found that the optimum pH for shoot proliferation and elongation was 5.5, which was severely inhibited in more acidic medium. In chickpea, pH 6.5 was proved to be the optimum for embryo formation, which was adversely affected by pH above 7.0 and below 4.0 (Barn and Wakhlu 1993). Wang et al. (2005) showed that pH level of 5.8–6.6 was broadly effective for shoot regeneration for *Camptotheca acuminata*. The best result of shoot regeneration was found for the medium at pH 5.8 with 90 % regeneration frequency. On a medium with pH 7.0 or below 5.4, the regenerated shoot number was low. Moreover, on the medium with pH value below 5.4, the regenerated shoots showed serious vitrification.

A similar study was conducted by Faisal et al. (2006a and b) where a wide range of pH was tested for shoot induction and it reported maximum multiplication rate at a pH of 5.8. Similar findings were reported by Siddique and Anis (2007a) in *Cassia angustifolia and B. monnieri* (Naik et al. 2010) and Perveen et al. (2011) in *Albizia lebbeck*. These studies indicated that the effect of pH value on plant regeneration depends on plant species.

2.4.4 Basal Media

The basic components of plant tissue culture media are the mineral nutrients. How rapidly a tissue grows and the extent and quality of morphogenetic responses are strongly influenced by the type and concentration of nutrients supplied.

2.4 Factors Affecting In Vitro Shoot Regeneration and Growth of Plants

Early research by Gautheret (1939); Heller (1953); White (1942); Hildebrandt et al. (1946); Nitsch and Nitsch (1956) culminated in the development of MS medium by Murashige and Skoog (1962). The organic and mineral compositions of the culture medium are particularly important to improve differentiation and to optimize explant's growth. It is well known that the amount of nutrients present in the culture medium must be sufficient to foster growth throughout the entire culture period.

The potential benefits of optimizing the nutrient component of culture media for a particular response are well documented across a wide range of species and applications. For example, the concentration of NH^{4+} and NO^{3-} affects numerous in vitro responses including the development of somatic embryos (Meijer and Brown 1987; Poddar et al. 1997; Elkonin and Pakhomova 2000; Leljak-Levanic et al. 2004), the plating efficiency of protoplasts (Attree et al. 1989), the efficiency of plant recovery after ovule culture (McCoy and Smith 1986), shoot regeneration (Leblay et al. 1991), regulation of growth and biomass of bioreactor-grown plantlets (Sivakumar et al. 2005), and controlling the rate of root initiation on shoot cultures (Hyndman et al. 1982).

Because there are 13 mineral elements essential for plant growth (Epstein and Bloom 2005), the experimental determination of optimal nutrient levels is complex. This complexity illustrates why the "revised medium" developed by Murashige and Skoog (1962) was an important development. Although MS medium is not optimal for many tissues, many tissues will grow on it to some degree; hence, MS medium represents a starting point to begin the process of improving a response. The significant and distinguishing feature of MS (1962) medium is its high nitrate, ammonium, and potassium contents.

The composition, type, and strength of basal medium also played an important role in shoot multiplication. Full strength of MS medium was found favorable for multiple shoot production in *H. antidysentrica* (Mallikarjuna and Rajendrudu 2007), *Actinidia deliciosa* (Akbas et al. 2007), *P. santalinus* (Rajeswari and Paliwal 2008), *Acacia nilotica* (Abbas et al. 2010), *Albizia lebbeck* (Perveen et al. 2011), and *V. negundo* (Ahmad and Anis 2011). Modification in the MS medium such as MS salts reduced to one half, one third, one fifth, or three fourth has been found effective in *Acacia senegal* (Badji et al. 1993), *A. mearnsii* (Huang et al. 1994), *Anacardium occidentale* (Das et al. 1996).

While many studies have concentrated on the influence of PGRs, the influence of the nutrient medium has received less attention. Most studies have used MS medium (Murashige and Skoog 1962) without modification, but a few have reduced the overall ion concentration or modified the nitrogen concentration or nitrogen sources (Chevreau et al. 1992; Bell and Reed 2002). Only a few of the studies made comparison among several nutrient media for axillary, adventitious shoot regeneration, as well as for somatic embryogenesis. Nedelcheva (1986) found that shoot proliferation of "Bartlett" was the greatest on a medium devised by Quoirin and Lepoivre (1977; QL), in comparison to MS medium. In contrast, Baviera et al. (1989) obtained better shoot proliferation of "Conference" on MS than on QL. Wang (1991) observed a higher degree of multiple shoot formation of the *Pyrus communis* L. rootstock BP10030 on woody plant medium (WPM; Lloyd and McCown 1981) and QL than on MS in a double-phase culture system consisting of a liquid medium overlaid on a semisolid medium. Yeo and Reed (1995) found that the nutrient medium of Cheng (1979) was better for shoot multiplication than WPM for a genetically diverse group of root stocks, such as "OH × F 230" (*P. communis* L.), "OPR 260" (*Pyrus betulifolia* Bunge), and "OPR 157" (*Pyrus calleryana* Decne), whereas Thakur and Kanwar (2008) reported that WPM resulted in enhanced axillary shoot proliferation of *Pyrus pyrifolia* when compared to MS and various modifications, and Gonzalez-Rodriguez et al. (2010) found that it was true for adventitious shoot regeneration from stem explants of *Tabebuia donnell-smithii* rose.

Recently, Khan et al. (2011) demonstrated the best shoot induction response in nodal explants of *Salix tetrasperma* in WPM medium supplemented with different PGRs. Bhatt and Dhar (2004) established a higher efficiency of WPM over other types of medium like B_5 (Gamborg et al. 1968) and MS and their different strengths were tried for shoot proliferation in *Myrica esculenta*. They observed that neither MS nor B_5 gave satisfactory response even when the salt concentration is reduced to half, and shoot response was severely inhibited.

Earlier, Gharyal and Maheshwari (1990) reported 36% direct shoot regeneration from petioles of different leguminous plant species on a B5 medium supplemented with 0.5 mg l^{-1} IAA and 1 mg l^{-1} BA. Lakshmisita et al. (1992) and Anuradha and Pulliah (1999) used B_5 medium for *P. marsupium* and *P. santilinus*. Similarly, Kaneda et al. (1997) found that B_5 medium was more superior over L_2 (Philips and Collins 1979) and MS media in obtaining maximum number of multiple shoots in *Glycine max*. Likewise, Wang et al. (2005) showed that B_5 and WPM media were the optimal basal media for shoot regeneration from axillary bud in *Camptotheca acuminata*, while Douglas and McNamara (2000) obtained adventitious shoot regeneration in *Acacia mangium* using Driver and Kuniyuki (1984) and McGranahan et al. (1987) (DKW) medium.

Bosela and Michler (2008) tested different media types in *Juglans nigra* where both MS and DKW media were suitable for long-term culture maintenance but the hyperhydricity frequencies were unacceptably high (70–100%) for the WPM and ½ × DKW nutrient formulations. The major differences in macronutrients among these media are in ammonium and nitrate ion concentrations and total ion concentration. Full-strength MS is high in ammonium (20.6 mM) and nitrate (39.4 mM) ions, while QL is a low-ammonium medium (5 mM). WPM contains low concentrations of both ammonium (5 mM) and nitrate (9.7 mM) ions. In addition, QL uses calcium nitrate as a nitrogen source. A medium originally developed for walnut (Driver and Kuniyuki 1984), designated DKW, also has lower ammonium ion content (17.7 mM) than MS and contains calcium nitrate instead of potassium nitrate.

2.4.5 Carbohydrate Source

Plants growing under tissue culture conditions are semiautotrophic (Hazarika 2003) and leaves formed during in vitro growth may never attain photosynthetic competence (Van Huylenbroeck and Debergh 1996). Moreover, plantlets growing under in vitro conditions have limited accessibility to CO_2 inside the vessel (Hazarika 2003). Therefore, sugar is supplemented as a carbon source to maintain an adequate supply of carbon source for in vitro multiplication and growth of plant cell, tissue, and organs or whole plantlets.

Addition of sugar to the culture media also helps in the maintenance of osmotic potential of cells and conservation of water (Hazarika 2003). The conservation of water is essentially important for ex vitro settlement of plants, because in vitro-grown plants lack a well-developed cuticle and epicuticular wax (water housekeeping system; Van Huylenbroeck et al. 2000). Moreover, exogenous supply of sugar increases starch and sucrose reserves in micropropagated plants and could favor ex vitro acclimatization and speed up physiological adaptations (Pospisilova et al. 1999a). Nonetheless, addition of sugar to the culture media has been shown to be negatively correlated with growth (Kwa et al. 1995), photosynthesis (Serret et al. 1997; Hazarika 2003), and expression of enzymes of the carbon assimilatory pathway (Kilb et al. 1996).

George and Sherrington (1984) had drawn the conclusion that for optimal growth and multiplication, 2–4% sucrose was found to be optimum whereas Capellades et al. (1991) found that the size and number of starch granules of Rosa cultivated in vitro increase with the sucrose level in the culture medium.

Hazarika et al. (2000) have demonstrated that in vitro preconditioning of *Citrus* microshoots with 3% sucrose concentration is advantageous for ex vitro survival and acclimatization. There was a linear increase in biochemical constituents, viz. reducing sugar, starch, and total chlorophyll on addition of sucrose to the medium. Likewise, high-frequency in vitro shoot multiplication of *Plumbago indica* was possible in a medium containing 3% sucrose (Chetia and Handique 2000).

Mehta et al. (2000) reported that an increase in sucrose concentration from 2 to 4% in the medium increases caulogenic response in tamarind plantlets from 34 to 48% in explants. On the contrary, studying acclimatization of Asiatic hybrid lily under stress conditions after propagation through tissue culture, Mishra and Dutta (2001) reported that liquid medium having 9% sucrose and other phytohormones was found suitable for the growth of bulblets in the isolated unrooted shoots. Due to the high concentration of sucrose, the size of the bulblets increased from 0.5 cm in diameter to approximately 1–1.5 cm within 2 months of inoculation. Further, the increase in sucrose to 6% induced browning of media which was detrimental for the growth of the shoots whereas the induction of multiple shoots using shoot tips of *Gerbera* was accomplished on MS medium supplemented with 3% sucrose and other phytohormones, and almost 100% survival rate was obtained after transfer (Ashwath and Choudhury 2002).

Akbas et al. (2007) tested the effects of different carbon sources (sucrose, maltose, and dextrose at 3% concentration) in *Actinidia deliciosa* and the best results were obtained on MS medium using 3% sucrose.

Recently, Jo et al. (2009) observed the effect of sucrose concentrations (0–9%) on the multiplication and growth of the plantlets of *Alocasia amazonica* under in vitro condition and their subsequent acclimatization under ex vitro condition. An absence of sucrose in the growth medium induced generation of leaves; however, it decreased multiplication.

Sucrose supply of 6 or 9% increased multiplication, corm size, fresh weight, dry weight, and root number; however, it decreased photoautotrophic growth (leaves). A similar reduction in the growth of plantlets was observed by Serret et al. (1997), when they incorporated sugar into the medium. It has previously been shown that decrease in sucrose concentration in the medium enhanced the photosynthetic ability of plantlets (Desjardins et al. 1995). Plantlets growing on the sucrose-supplemented media exhibit reduced photosynthesis, probably, due to the presence of sufficient energy source (sugars) for other metabolic activities (Rolland et al. 2002; Amiard et al. 2005). When in vitro-grown plantlets were transferred to ex vitro condition, the best growth was observed in those plantlets which have been micropropagated with 3% sucrose.

2.4.6 Subculture Passages

Subculturing often becomes imperative when the density of cells, tissues, or organs becomes excessive and when there is a need to increase the volume of a culture or to increase the number of organs (e.g., shoots or somatic embryos) for micropropagation.

Rapid rates of plant propagation depend on the ability to subculture shoots from proliferating shoot or node cultures, from cultures giving direct shoot regeneration, or from callus or suspensions capable of reliable shoot or embryo regeneration. The reason for transfer or subculture is that the growth of plant material in a closed vessel eventually leads to the accumulation of toxic metabolites and the exhaustion of the medium, or to its drying out. Thus, even to maintain the culture, all or part of it must be transferred onto fresh medium.

Shoot cultures are subcultured by segmenting individual shoots or shoot clusters. Bajaj et al. (1988) obtained around 2,200 plantlets of *Thymus vulgaris* from a single shoot grown in vitro in 5 months (four passages). Ajithkumar and Seeni (1998) reported that repeated subculturing of nodes and leaf from shoot cultures of *Aegle marmelos* helped to achieve a continuous production of callus-free healthy shoots at least upto five subculture cycles. Borthakur et al. (1999) established a mass multiplication protocol for *Alpinia galanga* by subculturing the regenerated explants to Kn-supplemented medium for more than 1 year. They obtained an average of 1,000 plantlets with four to five successive subculture cycles, i.e., within 40–45 days. Likewise, Raghu et al. (2007) observed that the micropropagated shoots of *Aegle marmelos* could be subcultured up to 20 cycles without loss of vigor to produce shoots free from morphological and growth abnormalities. Similarly, in *Simmondsia chinensis*, around 10–15 shoots were produced by repeated subculturing up to three successive subcultures (Singh et al. 2008).

The increase in shoot number may be due to the suppression of apical dominance during subculture that induced basal dormant meristematic cells to form new shoots (Shukla et al. 2009).

Hence, by adopting this procedure of shoot excision and reculturing of the mother explants in the fresh medium, a large number of shoots could be obtained per explants within few months (Asthana et al. 2011). This approach of increasing the yield of shoots at an enhanced pace was adopted earlier for other woody taxa (Kaveriappa et al. 1997; Jain and Babbar 2000; Hiregoudar et al. 2005; Prakash et al. 2006; Anis et al. 2010; Tripathi and Kumar 2010; Shekhawat and Shekhawat 2011).

2.5 Rooting of In Vitro-Regenerated Shoots

Rooting of in vitro-regenerated shoots and transplantation of the plantlets to the field is the most important, crucial, and essential step, but a difficult task in tissue culture of woody trees (Murashige 1974). Generally, rooting in micropropagated shoots can be achieved by two different methods, i.e., in vitro and ex vitro methods. Root induction and elongation are complex processes that are influenced by a large number of factors, such as genotype, type, and concentration of PGRs, and culture conditions (Bennett et al. 1994; Mylona and Dolan 2002).

The intricacies involved in adventitious rooting were reviewed by Haissig (1974), George and Sherrington (1984), Gaspar et al. (1994), and Rout et al. (2000). The in vitro-regenerated shoots of various medicinal plants rooted readily on growth regulator-free MS basal medium (Cristina et al. 1990; Binh et al. 1990; Faisal and Anis 2003; Shan et al. 2005; Mallikarjuna and Rajendrudu 2007; Jahan and Anis 2009). The ease with which microshoots root in vitro in the absence of exogenously supplied hormones was supported by the fact that there may occur high endogenous auxin in these in vitro-regenerated shoots. A full-strength MS medium was found to be satisfactory for root induction in a number of plant species but as the concentration of salts were reduced to half or to much lower levels (1/3 or 1/4), a striking increase in the rooting efficiency was observed. In woody trees, Rai et al. (2010) and Tripathi and Kumar (2010) have supported the fact that relatively low salt concentration in the medium is known to enhance rooting efficiency of microshoots. Recently, Phulwaria et al. (2012) have reported that half-strength MS medium supplemented with auxin was the most effective for rooting of shoots in *Terminalia bellirica*.

Exogenous auxins are often used in a number of plant species to promote in vitro rooting of in vitro-produced microshoots and their efficacy depends on several factors such as the affinity for the auxin receptor protein involved in rooting, the concentration of free auxin that reaches target competent cells, the amount of endogenous auxins, and the metabolic stability (De Klerk et al. 1999; Gaba 2005).

Auxins most frequently incorporated in the medium to induce rooting are IAA, IBA, and NAA. In particular, higher auxin concentrations are generally required during the induction phase, whereas during the formation phase growth regulators become inhibitory. This effect is particularly evident when microcuttings are cultured continuously on media with auxins, as demonstrated by De Klerk et al. (1997) in apple microcuttings previously exposed for 3 weeks to IAA and which rooted ex vitro more efficiently than those exposed to IBA or NAA. The different response was attributed to the higher stability and consequent supraoptimal concentrations of IBA and NAA for further rooting formation as compared to IAA, which, being unstable and easily degradable, would, instead, favor root growth and elongation. Nair and Seeni (2001) reported that MS medium supplemented with IAA, IBA, and NAA, for which IAA under initial dark conditions, gave the maximum rooting within a period of 5 weeks in *Celastrus paniculatus*.

Generally, IBA has been observed to induce strong rooting response over IAA or NAA and has been extensively used to promote rooting in a wide range of tree species such as *Myrtus communis* (Ruffoni et al. 1994; Scarpa et al. 2000), *Pistacia vera* (Onay 2000), *Terminalia arjuna* (Pandey and Jaiswal 2002; Pandey et al. 2006), *Terminalia chebula* (Shyamkumar 2003), *Kigelia pinnata* (Thomas and Puthur 2004), *Stereospermum personatum* (Shukla et al. 2009), *Semecar-*

pus anacardium (Panda and Hazra 2010), and *Sapindus trifoliatus* (Asthana et al. 2011). Moreover, this group of plant hormone is readily and easily available commercially worldwide (Epstein and Ludwig-Miller 1993). Ludwig-Muller (2000) has documented that the stimulatory effects of IBA on the root development may be due to several factors such as its preferential uptake, transport, and stability over other auxins and subsequent gene activation.

Many researchers emphasized on ex vitro rooting because plants developed after ex vitro rooting have a better root system than the plants raised after in vitro rooting (Borkowska 2001). In addition, the ex vitro technique is comparatively less time consuming and cost-effective and requires less labor, chemicals, and equipment than in vitro rooting because plantlets rooted ex vitro do not need any additional acclimatization prior to transplanting in the field conditions (Yan et al. 2010). The main advantages of ex vitro rooting are that the chance of root damage is less, rooting rates are good, and root quality is better (Bellamine et al. 1998).

Ex vitro rooting of in vitro-multiplied shoots has been reported for several species, including *G. jasminoides* (Economou and Spanoudaki 1985), apple (Zimmerman and Fordham 1985; Stimart and Harbage 1993), *Actinidia deliciosa* (Pedroso et al. 1992), *Cornus nuttallii* (Edson et al. 1994), blueberry (Isutsa et al. 1994), hazelnut (Nas and Read 2004), *Nyctanthes arbortristis* (Siddique et al. 2006), *V. negundo* (Ahmad and Anis 2007), *Malus zumi* (Xu et al. 2008), *Metrosideros excelsa* (Iapichino and Airo 2008), and *Terminalia bellirica* (Phulwaria et al. 2012). The system is now preferred in woody plant micropropagation to rapidly produce high-quality plantlets and to avoid the potential of off-types (Suttle 2000).

2.6 Acclimatization and Hardening of Plantlets

The ultimate success of in vitro propagation in a reforestation program depends on a reliable acclimatization protocol, ensuring low cost and high survival rates. In vitro protocols provide minimal stress and optimum conditions for shoot/plant multiplication (Hazarika 2006). As a consequence of these special conditions (e.g., high air humidity, low irradiance, low CO_2 during photoperiod, high levels of sugars as carbon source and growth regulators), in vitro-grown plantlets usually exhibit abnormal morphology, anatomy, and/or physiology (Pospisilova et al. 1999b; Premkumar et al. 2001; Hazarika 2006). Under these conditions, in vitro plantlets can develop specific features (e.g., nonfunctional roots and/or stomata) that are inconsistent with the development under greenhouse or field conditions. Also, the mixo-heterotrophic mode of nutrition and poor mechanism to control water loss render microphagated plants vulnerable to the transplantation shocks when directly placed in a greenhouse or field.

Understanding the physiological characteristics of micropropagated plants and the changes they undergo during the hardening process should facilitate the development of efficient transplantation protocols and will help to make decisions on, if necessary, adjusting environmental conditions (Hazarika 2006). For example, water/osmotic stress is often the cause of micropropagated plant's mortality and its monitorization is particularly important when acclimatization occurs in a degraded land as is the case reported by Brito et al. (2003).

The improvement of the photosynthetic competence during acclimation is a common characteristic in various plant species grown in vitro (Yue et al. 1993; Van Huylenbroeck et al. 1998a; Pospisilova et al. 1999b). Grout (1988) suggested that based on the behavior of in vitro-formed leaves, plants can be classified into two groups. In the first group, the in vitro leaves are photosynthetically competent and function as normally formed leaves; in the second group, these leaves act as storage organs and never become fully autotrophic. Moreover, the correlation between their performances and in vitro culture condition was also reported (Van Huylenbroeck et al. 1996, 1998a).

The presence of sugars in the medium may promote mixotrophy, leading to a downregulation of photosynthesis due to feedback inhibition of the Calvin cycle (Amancio et al. 1999; Premkumar et al. 2001; Van Huylenbroeck et al. 2000). Among photosynthetic enzymes, Rubisco deserves much attention since it performs a dual role as a catalyst in the carboxylation of CO_2 and as a major storage protein being 40–80 % of the total soluble leaf proteins (Premkumar et al. 2001). Both roles could be important in overcoming the critical acclimatization phase, when the mixo-heterotrophic behavior of the in vitro plants is shifted to an autotrophic functioning.

In vitro plantlets grow generally under low level of light, with plenty of sugar and nutrients to favor heterotrophic growth and in an atmosphere with high percentage of relative humidity (Hazarika 2003). Due to these factors, in vitro plants have low rates of photosynthesis and an incipient photosynthetic apparatus. After transfer to ex vitro conditions, most micropropagated plants develop a functional photosynthetic apparatus, although the increase in light intensity is not linearly translated to an increase in photosynthesis (Amancio et al. 1999). The results on the enhancement in the contents of the photosynthetic pigments in micropropagated plantlets during the acclimatization have been reported by Kadlecek et al. (1998) and Pospisilova et al. (1999b). The significant increase in chlorophyll (a and b) contents with exposure to high light levels suggested that chlorophyll synthesis enzymes vital for chlorophyll biosynthesis were induced. Since, both chlorophyll a and b are associated with light-harvesting complexes of photosystem II (PS II), an increase in the levels of pigments during the exposure to high light intensity did not impair the core complexes of PS II (Von Willert et al. 1995). Also, carotenoids play a key function in protecting chlorophyll pigments under stress conditions (Kenneth et al. 2000).

There are reports available where an initial abrupt decrease in chlorophyll contents during the starting days followed by a continuous and subsequent increase was noticed as in *Ocimum basilicum* (Siddique and Anis 2008) towards the final days of acclimatization. A similar pattern of photosynthetic efficiency in micropropagated plants of neem was detected by Lavanya et al. (2009). Faisal and Anis (2010) compared the chlorophyll contents of ex vitro-formed leaves of *Tylophora indica* with that of in vitro ones during the acclimation period and found significant higher levels of pigments in the fully hardened plantlets at the 28th day of acclimation.

Amancio et al. (1999) indicated that high light regime during acclimatization has a direct influence on the transition to in vitro characteristics and on the final yield, without symptoms of light stress. Also, it has been documented in the majority of the reports that the acclimatized plants had normal leaf development and lacked detectable morphological variation and showed apparently uniform growth and true-to-type morphology.

2.7 Advancement in Plant Tissue Culture: Synthetic Seed Technology

Synthetic seed technology in the last decade has emerged as one of the major branches of plant biotechnology and has opened up unprecedented opportunities in many areas of basic and applied biological researches. Tissue cultures of several plant species produce somatic embryos which proceed through similar developmental stages like that of seed zygotic embryos to form a plant. Synthetic seed refers to the encapsulation of somatic embryos or encapsulated buds, bulbs, or any form of meristem which can develop into plantlets.

Encapsulation is usually done in a suitable gel (sodium alginate) matrix to produce a "synthetic seed coat" and the resulting encapsulated propagules can be treated like natural seeds. The facility to incorporate nutrients, biofertilizers, antibiotics, or other essential additives to the matrix and the easy handling, storing, shipping, and planting are the major attractions to imply the synthetic seed as a unit of delivery of tissue-cultured plants.

The uniform and simultaneous production of somatic embryos in culture followed by uniform germination of encapsulated embryos could possibly remove many disadvantages associated with natural seeds. Many trees produce seeds in certain periods of the year whereas synthetic seeds would be available throughout the year.

The first synseeds of desiccated carrot and celery somatic embryos (Kitto and Janick 1982) were obtained on encapsulation in "Polyox," polyoxyethylene, which is soluble in water, nontoxic to embryos, inert to microbial growth, and dries to form a thin film.

Synseeds of alfalfa were obtained on encapsulation of somatic embryos in 2% solution of calcium alginate (Redenbaugh et al. 1991). Since then, this has been the most practiced method of synseed production. An embryo and sodium alginate mixture was dropped in 100 mM solution of $CaCl_2 \cdot 2H_2O$. In an ion exchange reaction, Na^+ ions were replaced by Ca^{2+} ions forming calcium alginate gel. For the production of synseed of carrot, 1% Na-alginate, 50 mM Ca^{2+}, and 20–30 min of complexing were satisfactory (Molle et al. 1993).

The technique of synseed has been widely studied and works with various plant species including fruits, cereals, medicinal plants, vegetables, ornamentals, forest trees, and orchids (Ballester et al. 1997; Gonzalez-Benito et al. 1997; Standardi and Piccioni 1998; Pattnaik and Chand 2000; Ara et al. 2000; Nyende et al. 2003; Chand and Singh 2004; Tsvetkov and Hausman 2005; Malabadi and Staden 2005; Naik and Chand 2006; Faisal et al. 2006; Faisal and Anis 2007; Antonietta et al. 2007; Micheli et al. 2007; Pintos et al. 2008; Rai et al. 2008; Singh et al. 2009; Sundararaj et al. 2010; Singh et al. 2010; Germana et al. 2011; Mishra et al. 2011). In addition, during cold storage, encapsulated nodal segments require no transfer to fresh medium, thus reducing the cost of maintaining germplasm cultures (West et al. 2006).

In this context, the most important application of synthetic seeds for these plants could be in exchange of elite and axenic plant material between laboratories due to small bead size and relative ease of handling these structures (Rai et al. 2008).

Conservation is an important aspect of encapsulation technology. For short- and medium-term storage, the aim is to increase the interval between subcultures by reducing growth. This is achieved by modifying the environmental conditions and/or the culture medium. Various approaches have been applied for slow-growth maintenance of cultures such as maintenance under reduced temperature and/or reduced light intensity, use of growth retardants such as ABA, use of minimal growth medium (restrict the growth of cultures by alteration of mineral content and/or sucrose in medium), use of osmoticum, reduction in oxygen concentration, and combination of more than one treatment (Gupta and Mandal 2003).

Long-term storage of synthetic seeds can be achieved through storage at ultralow temperature, termed as cryopreservation, and is usually carried out by using liquid nitrogen ($-196\ °C$). At this temperature, all cellular division and metabolic processes are suspended, and hence plant material can be stored for an unlimited period (Engelmann 1997). In recent years, several new cryopreservation techniques like encapsulation-dehydration and encapsulation-vitrification methods based on technology have been developed for the production of synseeds (Fig. 2.1).

2.8 Clonal Fidelity of Micropropagated Plants

Clonal fidelity is a major consideration in commercial micropropagation using in vitro tissue culture methods. In vitro-regenerated plants are usually susceptible to genetic changes due to culture stress and might exhibit somaclonal variation (Larkin and Scowcroft 1981). Somaclonal variation was first reported for woody plants in *Citrus grandis* (Chaturvedi and Mitra 1975). Therefore, it should be a common practice to assure the trueness of tissue culture plants after regeneration.

However, somaclonal variation is of immense importance for the isolation of improved clones of forest trees (Ahuja 1993). In addition to somaclonal variation, mislabeling and mixing of clones in germplasm collections have also been reported (Keil and Griffin, 1994). Materials from germplasm banks are frequently used in breeding and tree improvement programs. The economic implications of such inadvertent variations and mixings of accessions could be serious as considerable time and money are spent before the mistakes are detected. This necessitates the development of suitable strategies for assessing genetic uniformity and for identifying the variations.

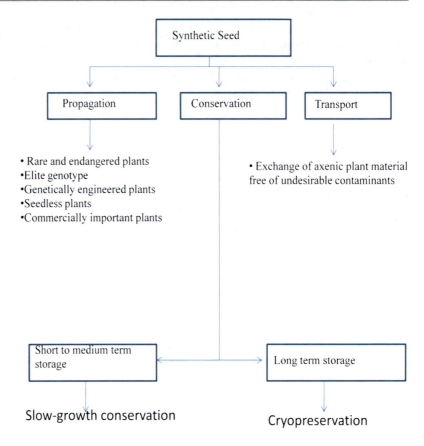

Fig. 2.1 Scope of synthetic seeds. (Source: Rai et al. 2009)

The need to test the genetic fidelity of tissue culture plants in tree species is important because they are harvested on long rotations and the in vitro cultures are maintained through many subcultures. Molecular techniques are at present powerful and valuable tools in the analysis of genetic fidelity of in vitro-propagated plants. Evaluation of genetic diversity using molecular techniques could provide useful baseline information for breeding programs.

Molecular markers are not influenced by environmental factors and can be estimated using DNA from any growth stage and, therefore, are very useful in assessing genetic diversity of plant species. Among the various molecular markers, polymerase chain reaction (PCR)-based markers, such as random amplified polymorphic DNA (RAPD; Williams et al. 1990), inter-simple sequence repeat (ISSR; Zietkiewiez et al. 1994), sequence-related amplified fragment length polymorphism (Li and Qulros 2001), and amplified fragment length polymorphism (AFLP; Vos et al. 1995), have become popular as their application does not require any prior sequence information.

2.8.1 PCR-Based DNA Markers

The development of PCR for amplifying DNA sequences led to the revolution in the applicability of molecular methods, and a range of new technologies were developed which could overcome the technical limitations of hybridization-based methods. In a PCR, arbitrary or known sequence primers are used to amplify one or discrete DNA segments that can be resolved in agarose or polyacrylamide gels. Each product is derived from a region of the genome containing two DNA sites with sequences complementary to the primer(s) on the opposite strand and sufficiently close for the amplification to work.

2.8.1.1 RAPD-PCR Markers

Welsh and McClelland (1991) developed a new PCR-based genetic assay, namely RAPD. This procedure detects nucleotide sequence polymorphisms in DNA by using primer of arbitrary nucleotide sequence. In this reaction, a single species of primer anneals to the genomic DNA at two different sites on the complementary strands of the DNA template. If these primary sites are within an amplifiable range of each other, a discrete DNA product is formed through thermocyclic amplification. On an average, each primer directs amplification of several discrete loci in the genome, making the assay useful for efficient screening of nucleotide sequence polymorphism among individuals.

Polymorphisms in RAPD result from different types of changes in the genomic DNA: base-pair substitution, insertions, and deletions, which modify or eliminate the primer-annealing sites; insertions in the genomic sequence that change the intervening length of DNA between the primer sites; and insertions which separate the primer sites to a distance that will not permit amplifications (Williams et al. 1990). Each of these results in the presence or absence of a particular RAPD fragment. This procedure usually amplifies 1–15 DNA fragments from a single primer reaction (Reiter et al. 1992). The primers are usually 10 bp in length with GC content of at least 50 % and have a low annealing temperature (36–40°C). However, due to the stochastic nature of DNA amplification with random sequence primers, it is important to optimize and maintain consistent reaction conditions for reproducible DNA amplification. RAPD is the method generally employed for the detection of genetic diversity because it has the advantage of being technically simple, quick to perform, and requires only small amounts of DNA (Ceasar et al. 2010).

Many investigators have reported genetic stability in *Picea mariana* (Isabel et al. 1993), *Pinus thunbergii* (Goto et al. 1998), *Lilium* (Varshney et al. 2001), chestnut rootstock hybrid (Carvalho et al. 2004), *Prunus dulcis* (Martins et al. 2004), banana (Venkatachalam et al. 2007), *V. negundo* (Ahmad and Anis 2011), *Sapindus trifoliatus* (Asthana et al. 2011), *P. santalinus* (Balaraju et al. 2011), *Simmondsia chinensis* (Kumar et al. 2011), etc. using RAPD.

2.8.1.2 ISSR-PCR Markers

In this technique, primers based on microsatellites are utilized to amplify ISSR DNA sequences. Here, various microsatellites anchored at the 3′ or 5′ end are used for amplifying genomic DNA which increases their specificity. This technique is more reproducible and generates three to five times the variation of RAPD (bands/marker; Nagaoka and Ogihara 1997). ISSR markers have been shown to be more reliable and conform closely to dominant Mendelian inheritance which makes them useful for genotype analysis and genome mapping (Tsumara et al. 1996; Nagaoka and Ogihara 1997). An unlimited number of primers can be synthesized for various combinations of di-, tri-, tetra-, penta-nucleotides $[(4)^3 = 64, (4)^4 = 256]$, etc. with an anchor made up of a few bases and can be exploited for a broad range of applications in plant species.

ISSR markers have been successfully applied in the analysis of genetic fidelity within lines of cauliflower (Leroy et al. 2000), almond (Martins et al. 2004), banana (Venkatachalam et al. 2007), *Swertia chirayita* (Joshi and Dhawan 2007), *Dictyospermum ovalifolium* (Chandrika et al. 2008), *Platanus acerifolia* (Huang et al. 2009), and *Nothapodytes foetida* (Chandrika and Rai 2010). Moreover, ISSR markers offer other advantages in the detection of somaclonal variation, notably a high degree of sensitivity, reproducibility, and the dominant representation of polymorphic genetic alleles.

2.9 Antioxidant Enzymes

It is known that reactive oxygen species (ROS) production increases during abiotic and biotic stresses and that ROS and some resultant metabolites are important signaling molecules (Moller et al. 2007). According to these authors, the production of ROS, such as superoxide (O_2^-), hydrogen peroxide (H_2O_2), hydroxyl radicals (HO^-), and singlet oxygen ($^1O_2$), is a consequence of aerobic metabolism, both as products of mainstream enzymatic reactions such as photorespiration or as an unavoidable accident (e.g., the O_2^- produced by the mitochondrial electron transport chain).

The chloroplasts, peroxisomes, and mitochondria are the main centers for ROS production in green plants, producing several types of ROS that have different properties and specific cellular roles (Moller et al. 2007). $^1O_2$ is produced in PS II and O_2^- at both photosystem I (PS I) and mitochondria, while the peroxisomes produce O_2^- and H_2O_2 in several key metabolic reactions (Moller et al. 2007). H_2O_2 is relatively stable and can be removed by normal cellular antioxidant systems (Yannarelli et al. 2006).

The other ROS are very unstable and are present at much lower concentrations. Several authors have reported oxidative damage caused by ROS and resultant products (Clijsters et al. 1999; Mittler 2002; Moller et al. 2007; Smeets et al. 2005; Ammar et al. 2008), including: (1) membrane damage caused by changes in the lipid composition of cellular membranes and accumulation of lipid peroxidation products, e.g., aldehydes such as malondialdehyde (MDA) and complex mixtures of lipid hydroperoxides; (2) oxidative damage of proteins leading to denaturation of functional and structural proteins, with inhibition of some enzymatic systems; (3) oxidative damage of nucleic acids; (4) oxidation of other components of the antioxidative system such as glutathione (GSH) and ascorbate; and (5) oxidation of free carbohydrates like sugars and polyols by reaction with HO^-.

According to Mittler (2002), plant cells require both the control of low levels of ROS for signaling purposes and the stress-induced detoxification of excess ROS. To help in the detoxification of excess ROS, plants have an efficient antioxidant defense system composed of enzymatic and nonenzymatic mechanisms (Gratao et al. 2005; Yannarelli et al. 2006) located in distinct cell organelles (peroxisomes, chloroplasts, and mitochondria). These enzymatic mechanisms include enzymes such as superoxide dismutase (SOD), catalase (CAT), ascorbate peroxidase (APX), and glutathione reductase (GR).

SOD catalyzes the dismutation of superoxide into H_2O_2 and O_2. SOD is found in almost all cellular compartments and is one of the main ROS-scavenging pathways of plants, participating in the water–water and the ascorbate–glutathione cycles in chloroplasts, cytosol, mitochondria, peroxisomes, and apoplast (Mittler 2002).

In plants, CAT is one of the main H_2O_2-scavenging enzymes that converts H_2O_2 into H_2O and O_2 in peroxisomes and is involved in the decomposition of H_2O_2 formed during photorespiration, without the need of an additional substrate (Pereira et al. 2002). The elimination of H_2O_2 in other cell compartments depends on distinct peroxidases, such as APX and glutathione peroxidase (GPX), which catalyze the breakdown of H_2O_2 using different reducing substrates (Pereira et al. 2002). APX isozymes are localized in chloroplasts, cytosol, mitochondria, peroxisomes, and apoplast (Jimenez et al. 1997; Mittler 2002). GPX is located in the cytosol (Mittler 2002) and also metabolizes H_2O_2, although at rates that are small compared with the large rates of H_2O_2 generation in plants (Noctor et al. 2002). Other peroxidases such as guaiacol peroxidases are enzymes that metabolize H_2O_2 to water, using the oxidation of a wide variety of substrates, mainly phenols. These enzymes are also involved in numerous physiological roles in plant tissues, including lignin biosynthesis and pathogen defense, among others (Yannarelli et al. 2006).

The ascorbate–glutathione cycle is also an important defense mechanism against oxidative stress caused by metals (Cuypers et al. 2002; Mittler 2002; Smeets et al. 2005). The different affinities of APX (μM range) and CAT (mM range) for H_2O_2 suggest that they belong to two different classes of H_2O_2-scavenging enzymes with CAT participating in the removal of excess ROS during stress (Mittler 2002).

The ubiquitous tripeptide GSH, which occurs mostly as a low molecular weight thiol compound in almost all cells, acts as a disulphide reductant to protect the thiol groups of enzymes, regenerate ascorbate, and react with $^1O_2$ and $OH^.$. GSH detoxifies herbicides by conjugation, either spontaneously or by the activity of a glutathione-S-transferase (GST), and regulates gene expression in response to environmental stress and pathogen attack. GSH also participates in the regeneration of ascorbate from docosahexaenoic acid (DHA) via the enzyme dehydroascorbate reductase (DHAR; Noctor and Foyer 1998). GR catalyses the NADPH-dependent formation of a disulphide bond in glutathione disulphide (GSSG) and is thus important for maintaining the reduced pool of GSH. The role of GSH and GR in H_2O_2 scavenging has been well established in the Halliwell–Asada pathway (Noctor and Foyer 1998; Asada 2000). GR catalyzes the rate-limiting last step of the Halliwell–Asada pathway. An increase in GR activity in plants results in the accumulation of GSH and ultimately confers stress tolerance in plants. Expression of GR is unregulated under stresses such as high light, mechanical wounding, high temperature, chilling, and exposure to heavy metals and herbicides (Apel and Hirt 2004; Karuppanapandian et al. 2011).

Van Huylenbroeck et al. (2000) reported that micropropagated plants develop antioxidant mechanism during acclimation. Under high irradiance, significant changes in the activity of the antioxidant enzymatic system were observed in micropropagated plants of *Calathea* (Van Huylenbroeck 2000), *Phalaenopsis* (Ali et al. 2005), *Zingiber officinale* (Guan et al. 2008), *Rauwolfia tetraphylla* (Faisal and Anis 2009), *Tylophora indica* (Faisal and Anis 2010), *Ocimum basilicum* (Siddique and Anis 2009b), *Tecomella undulata* (Varshney and Anis 2011), and *Ulmus minor* (Dias et al. 2011).

In addition, Batkova et al. (2008) recognized that ex vitro transfer is often stressful for in vitro-grown plantlets. Water stress and photoinhibition, often accompanying the acclimatization of in vitro-grown plantlets to ex vitro conditions, are probably the main factors promoting the production of ROS and, in consequence, oxidative stress. The extent of the damaging effects of ROS depends on the effectiveness of the antioxidative systems which include low molecular mass antioxidants (ascorbate, glutathione, tocopherols, carotenoids, phenols) and antioxidative enzymes (SOD, APX, CAT, GR, monodehydroascorbate reductase, DHAR). Authors have focused on ROS production and development of antioxidative system during in vitro growth and their further changes during ex vitro transfer.

References

Abbas H, Qaiser M, Naqvi B (2010) Rapid *in vitro* multiplication of *Acacia nilotica* subsp. *hemispherica*, a critically endangered endemic taxon. Pak J Bot 42:4087–4093

Adams AN (1972) An improved medium for strawberry meristem culture. J Hortic Sci 47:263–264

Ahee J, Duhoux E (1994) Root culturing of Faidherbia = *Acacia albida* as a source of explants for shoot regeneration. Plant Cell Tiss Org Cult 36:219–225

Ahmad N, Anis M (2011) An efficient *in vitro* process of recurrent production of cloned plants of *Vitex negundo* L. Eur J Fores Res 130:135–144

Ahuja MR (1993) Micropropagation a la carte In: Ahuja MR (ed) Micropropagation of woody plants, forestry series. Kluwer Academic, Netherlands 41, p 3–9

Ajithkumar D, Seeni S (1998) Rapid clonal multiplication through *in vitro* axillary shoot proliferation of *Aegle marmelos* (L.) Corr., a medicinal tree. Plant Cell Rep. 17:422–426.

Akbas FA, Isikalan C, Namli S, Basaran D (2007) Micropropagation of Kiwifruit (*Actinidia deliciosa*). Int J Agri Biol 9:389–493

Akin-Idowu PE, Ibitoye DO, Ademoyegun OT (2009) Tissue culture as a plant production technique for horticultural crops. Afri J Biotechnol 8:3782–3788

Ali MB, Hahn EJ, Paek KY (2005) Effects of light intensities on antioxidant enzymes and malondialdehyde content during short-term acclimatization on micropropagated *Phalaenopsis* plantlet. Environ Exp Bot 54:109–120

Amancio S, Rebordao JP, Chaves MM (1999) Improvement of acclimatization of micropropagated grapevine: Photosynthetic competence and carbon allocation. Plant Cell Tiss Org Cult 58:31–37

Amiard V, Mueh KE, Demmig-Adams B, Ebbert V, Turgeon R, Adams WW III (2005) Anatomical and photosynthetic acclimation to the light environment in species with differing mechanisms of phloem loading. Proc Natl Acad Sci USA 102:12968–12973

Ammar W, Nouairi I, Zarrouk M, Ghorbel M, Jemal F (2008) Antioxidative response to cadmium in roots and leaves of tomato plants. Biol Plant 52:727–731

Amo-Marco JB, Ibanez MR (1998) Micropropagation of *Limonium cavanillesii* Erben., a threatened statice, from inflorescence stems. Plant Growth Regul 24:49–54

Anis M, Ahmad N, Siddique I, Varshney A, Naz R, Perveen S, Khan Md I, Ahmed Md R, Husain MK, Khan PR, Aref IM (2012) Biotechnological approaches for the conservation of some forest tree species. In: Jenkins JA (ed) Forest decline: causes and impacts. Nova Publishers, Inc., p 1–39

Anis M, Varshney A, Siddique I (2010) *In vitro* clonal propagation of *Balanites aegyptiaca* (L.) Del. Agrofores Syst 78:151–158

Antonietta G, Ahmad H, Maurizio M, Alvare S (2007) Preliminary research on conversion of encapsulated somatic embryos of *Citrus reticulata* Blanco. cv. Mandarino Tardivo di Ciaculli. Plant Cell Tiss Org Cult 88:117–120

Anuradha M, Pullaih T (1999) In vitro seed culture and induction of enhanced axillary branching in *Pterocarpus santalinus* and a method for rapid multiplication. Phytomorph 49:157–163

Apel K, Hirt H (2004) Reactive oxygen species: metabolism, oxidative stress and signal transduction. Annu Rev Plant Biol 55:206–216

Ara H, Jaiswal U, Jaiswal VS (2000) Synthetic seed: prospects and limitations. Curr Sci 12:1438–1444

Arora K, Sharma M, Srivastava J, Ranade SA, Sharma AK (2010) In vitro cloning of *Azadirachta indica* from root explants. Biol Plant 55:164–168

Asada K (2000) The water-water cycle as alternative photon and electron sinks. Phil Trans R Soc Lond B Biol Sci 355:1419–1431

Ashwath CR, Choudhary ML (2002) Rapid plant regeneration from *Gerbera jamesonii* Bolus callus cultures. Acta Bot Croat 61:125–134

Asthana P, Jaiswal VS, Jaiswal U (2011) Micropropagation of *Sapindus trifoliatus* L. and assessment of genetic fidelity of micropropagated plants using RAPD analysis. Acta Physiol Plant 33:1821–1829

Attree SM, Dunstan DI, Fowke LC (1989) Initiation of embryogenic callus and suspension cultures, and improved embryo regeneration from protoplasts, of white spruce (*Picea glauca*). Can J Bot 67:1790–1795

Augustine AC, D'Souza L (1997) Micropropagation of an endangered forest tree *Zanthoxylum rhetsa* Roxb. Phytomorph 47:319–323

Badji S, Mairone Y, Merlin G, Danthu P, Neville P, Colonna JP (1993) In vitro propagation of the gum arabic tree (*Acacia senegal* (L.) Willd): 1. Developing a rapid method for producing plants. Plant Cell Rep 12:629–633

Bajaj YPS, Furmanowa M, Olszowska O (1988) Biotechnology of micropropagation of medicinal and aromatic plants. In: Bajaj YPS (ed) Biotechnology in agriculture and forestry medicinal plants Vol 4. Springer-Verlag, Berlin, p 60–103

Baker BS, Bhatia SK (1993) Factors effecting adventitious shoot regeneration from leaf explants of quince (*Cydonia oblonga*). Plant Cell Tiss Org Cult 35:273–277

Balaraju K, Agastian P, Ignacimuthu S, Park K (2011) A rapid *in vitro* propagation of red sanders (*Pterocarpus santalinus* L.) using shoot tip explants. Acta Physiol Plant 33:2501–2510

Ball EA (1946) Development in sterile culture of stem tips and subjacent regions of *Tropaeolum majus* L. and of *Lupinus albus* L. Am J Bot 33:301–318

Ballester A, Janeiro LV, Vietez AM (1997) Cold storage of shoot cultures and alginate encapsulation of shoot tips of *Camellia japonica* L. and *Camellia reticulata* Lindley. Sci Hort 71:67–78

Barciszewski J, Rattan SIS, Siboska G, Clark BFC (1999) Kinetin—45 years on. Plant Sci 148:37–45

Barn KS, Wakhlu AK (1993) Somatic embryogenesis in plant regeneration from callus cultures of chickpea (*Cicer arietinum* L. [J]). Plant Cell Rep 11:71–75

Bashir MA, Rashid H, Anjum MA (2007) In vitro shoot initiation from nodal explants of jojoba (*Simmondsia chinesis*) strains. Biotechnol 6:165–174

Batkova P, Pospisilova J, Synkova H (2008) Production of reactive oxygen species and development of antioxidative systems during in vitro growth and ex vitro transfer. Biol Plant 52:413–422

Baviera JA, Garcia JL, Ibarra M (1989) Commercial *in vitro* micropropagation of pear cv. Conference. Acta Hortic 256:63–68

Bellamine J, Penel C, Greppin H, Gaspar T (1998) Confirmation of the role of auxin and calcium in the late phases of adventitious root formation. Plant Growth Regul 26:191–194

Bell RL, Reed BM (2002) *In vitro* tissue culture of pear: Advances in techniques for micropropagation and germplasm preservation. Acta. Hortic. 596: 412–418

Bennett IJ, McComb JA, Tonkin CM, McDavid DAJ (1994) Alternating cytokinins in multiplication media stimulates *in vitro* shoot growth and rooting of *Eucalyptus globulus* Labill. Ann Bot 74:53–58

Benson EE (2000) *In vitro* recalcitrance: an introduction. Special symposium: *in vitro* plant recalcitrance. Vitro Cell Dev Biol-Plant 36:141–148

Bergmann BA, Moon HK (1997) *In vitro* adventitious shoot production in *Paulownia*. Plant Cell Rep. 16:315–319

Bhat SR, Chitralekha P, Chandel KPS (1992) Regeneration of plants from long-term root culture of lime, *Citrus aurantifolia* (Chirstm.) Swing. Plant Cell Tiss Org Cult 29:19–25

Bhatia P, Ashwath N (2005) Effect of medium pH on shoot regeneration from the cotyledonary explants of tomato. Biotechnol 4:7–10

Bhatt ID, Dhar U (2004) Factors controlling micropropagation of *Myrica esculenta* buch.-Ham. ex D. Don.: a high value wild edible of Kumaun Himalaya. Afri J Biotechnol 3:534–540

Bhattacharya S, Bhattacharya S (1997) Rapid multiplication of *Jasminium officinale* L. by *in vitro* culture of nodal explants. Plant Cell Tiss Org Cult 51:57–60

Bhau BS, Wakhlu AK (2001) Effect of genotype, explant type and growth regulators on organogenesis in *Morus alba*. Plant Cell Tiss Org Cult 66:25–29

Bhojwani SS (1980) *In vitro* propagation of garlic by shoot proliferation. Sci Hort 13:47–52

Bhuyan AK, Pattnaik S, Chand PK (1997) Micropropagation of curry leaf tree [*Murraya koenigii* (L.) Spring.] by axillary proliferation using intact seedlings. Plant Cell Rep 16:779–782

Binh LT, Muoi LT, Oanh HTK, Thang RD, Phong DT (1990) Rapid propagation of *Agave* by *in vitro* tissue culture. Plant Cell Tiss Org Cult 23:67–70

Blinstrubiene A, Sliesaravicius A, Burbulis N (2004) Factors affecting morphogenesis in tissue culture of linseed flax (*Linum usitatissimum* L.). Acta Universitatis Latviensis Biol 676:149–152

Boerjan W (2005) Biotechnology and the domestication of forest trees. Curr Opi Biotechnol 16:159–166

Borkowska B (2001) Morphological and physiological characteristics of micropropagated strawberry plants rooted *in vitro* or *ex vitro*. Sci Hortic 89:195–206

Borthakur M, Dutta K, Nath SC, Singh RS (2000) Micropropagation of *Eclipta alba* and *Eupatorium adenophorum* using a single step nodal cutting technique. Plant Cell Tiss Org Cult 62:239–242

Borthakur M, Hazarika J, Singh RS (1999) A protocol for micropropagation of *Alpinia galanga*. Plant Cell Tiss Org Cult 55:231–233

Bosela MJ, Michler CH (2008) Media effects on black walnut (*Juglans nigra* L.) shoot culture growth *in vitro*: evaluation of multiple nutrient formulations and cytokinin types. In Vitro Cell Dev Biol-Plant 44:316–329

Brito G, Costa A, Fonseca H, Santos C (2003) Response of *Olea europaea* ssp. maderensis *in vitro* shoots to osmotic stress. Sci Hortic 97:411–417

Butcher D, Street HE (1964) Excised root culture. Biol Rev 30:513

Butenko RG, Lipsky AKH, Chernyak ND, Arya HC (1984) Changes in culture medium pH by cell suspension cultures of *Dioscorea deltoides*. Plant Sci Lett 35:207–212

Campbell MM, Brunner AM, Jones HM, Strauss SH (2003) Forestry's fertile crescent: the application of biotechnology to forest trees. Plant Biotechnol 1:141–154.

Capellades M, Lemeur L, Debergh P (1991) Effects of sucrose on starch accumulation and rate of photosynthesis in *Rosa* cultured *in vitro*. Plant Cell Tiss Org Cult 25:21–26

Carvalho LC, Goulao L, Oliveira C, Goncalves JC, Amancio S (2004) RAPD assessment for identification of clonal identity and genetic stability of *in vitro* propagated chestnut hybrids. Plant Cell Tiss Org Cult 77:23–27

Ceasar SA, Maxwell SL, Prasad KB, Karthigan M, Ignacimuthu S (2010) Highly efficient shoot regeneration of *Bacopa monnieri* L. using a two-stage culture procedure and assessment of genetic integrity of micropropagated plants by RAPD. Acta Physiol Plant 32:443–452

Centeno ML, Rodriguez A, Feito I, Fernandez B (1996) Relationship between endogenous auxin and cytokinin levels and morphogenic responses in *Actinidia deliciosa* tissue cultures. Plant Cell Rep 16:58–62

Chan JT, Chang WC (2002) Effect of tissue culture conditions and explant characteristics on direct somatic embryogenesis in *Oncidium* 'Grower Ramsay'. Plant Cell Tiss Org Cult 69:41–44

Chand S, Pattnaik S, Chand PK (2002) Adventitious shoot organogenesis and plant regeneration from cotyledons of *Dalbergia sissoo* Roxb., a timber yielding tree legume. Plant Cell Tiss Org Cult 68:203–209

Chand S, Singh AK (2004) Plant regeneration from encapsulated nodal segments of *Dalbergia sisso* Roxb., a timber yielding leguminous tree species. J Plant Physiol 161:237–243

Chandrasekhar T, Hussain TM, Jayanand B (2005) *In vitro* micropropagation of *Boswellia ovalifoliolata*. Natl Cent Biotechnol Inf 60:505–507

Chandrika M, Rai RV, Thoyajaksha (2010) ISSR marker based analysis of micropropagated plantlets of *Nothapodytes foetida*. Biol Plant 54:561–565

Chandrika M, Thoyajaksha, Rai RV, Kini RK (2008) Assessment of genetic stability of *in vitro* grown *Dictyospermum ovalifolium*. Biol Plant 52:735–739

Chattopadhyay S, Datta SK, Mahato SB (1995) Rapid micropropagation for *Mucuna pruriens* f. *pruriens*. L. Plant Cell Rep 15:271–273

Chaturvedi HC, Mitra GC (1975) A shift in morphogenetic pattern in *Citrus* callus tissue during prolonged culture. Ann Bot 39:683–687

Chaturvedi HC, Sharma M, Sharma AK, Jain M, Agha BQ, Gupta P (2004) *In vitro* germplasm preservation through regenerative excised root culture for conservation of phytodiversity. Indian J Biotechnol 3:305–315

Chaudhary H, Sood N (2008) Purification and partial characterization of lectins from *in vitro* cultures of *Ricinus communis*. Plant Tiss Cult Biotechnol 18:89–102

Cheng TY (1979) Micropropagation of clonal fruit tree rootstocks. Compact. Fruit Tree 12:127–137

Chetia S, Handique PJ (2000) High frequency *in vitro* multiplication of *Plumbago indica*- A medicinal plant. Curr Sci 78:1187

Chevreau E, Thibault B, Arnaud Y (1992) Micropropagation of pear (*Pyrus communis* L.). In: Bajaj YPS. (ed) Biotechnology in agriculture and forestry, vol. 18. Springer, Berlin. p 224–261

Clijsters H, Cuypers A, Vangronsveld J (1999) Physiological responses to heavy metals in higher plants: defence against oxidative stress. Z Naturforsch 54:730–734

Cocking EC (1960) A method for the isolation of plant protoplasts and vacuoles. Nature 187:927–929

Coenen C, Lomax TL (1997) Auxin-cytokinin interactions in higher plants: old problems and new tools. Trends Plant Sci 2:351–356

Cohen D (1980) Application of micropropagation methods for blueberries and tamarillos. Comb. Proc. Int Plant Prop Soc 30:144–146

Corredoira E, Ballester A, Vieitez AM (2008) Thidiazuron-induced high-frequency plant regeneration from leaf explants of *Paulownia tomentosa* mature trees. Plant Cell Tiss Org Cult 95:197–208

Cortizo M, Cuesta C, Centeno ML, Rodriguez A, Fernandez B, Ordas R (2009) Benzyladenine metabolism and temporal competence of *Pinus pinea* cotyledons to form buds *in vitro*. J Plant Physiol 166:1069–1076

Cristina M, Dos Santos F, Esquibel MA, Dos Santos VP (1990) *In vitro* propagation of the alkaloid producing plant *Datura insignis* Barb. Rodr. Plant Cell Tiss Org Cult 21:57–61

Cuypers A, Vangronsveld J, Clijsters H (2002) Peroxidases in roots and primary leaves of *Phaseolus vulgaris* copper and zinc phytotoxicity: a comparison. J Plant Physiol 159:869–876

Dalal NV, Rai VR (2004) *In vitro* propagation of *Oroxylum indicum* Vent. a medicinally important forest tree. J For Res 9:61–65

Das DK, Shiv Prakash N, Bhalla-Sarin N (1999) Multiple shoot induction and plant regeneration in litchi (*Litchi chinensis* Sonn.). Plant Cell Rep 18:691–695

Das S, Jha TB, Jha S (1996) *In vitro* propagation of cashew nut. Plant Cell Rep 15:615–619

Davies PJ (2004) Regulatory factors in hormone action: Level, location and signal transduction. In: Davies PJ (ed) plant Hormones, Kluwer Academic Publishers, Dordrecht, p 16–35

De Fossard RA, De Fossard H (1988) Micropropagation of some members of the Myrtaceace. Acta Hortic 227:346–351

De Klerk GJ, Ter Brugge J, Marinova S (1997) Effectiveness of indole acetic acid, indolebutyric acid and naphthalene acetic acid during adventitious root formation *in vitro* in Malus 'Jork 9'. Plant Cell Tiss Org Cult 49:39–44

De Klerk GJ, Van Der Krieken W, De Jong JC (1999) The formation of adventitious roots: New concepts, new possibilities. In Vitro Cell Dev Biol- Plant 35:189–199

Desjardins Y, Hdider C, De Riek J (1995) Carbon nutrition *in vitro*-regulation and manipulation of carbon assimilation in micropropagated systems. In: Aitken-Christie J, Kozai T, Smith MLA (eds) Automation and environmental control in plant tissue culture. Kluwer Academic Publishers, Netherlands, p 441–471

Detrez C, Ndiaye S, Dreyfus B (1994) *In vitro* regeneration of the tropical multipurpose leguminous tree *Sesbania grandiflora* from cotyledon explants. Plant Cell Rep 14:87–93

Devi YS, Mukherjee BB, Gupta S (1994) Rapid cloning of elite teak (*Tectona grandis* L.) by *in vitro* multiple shoot production. Indian J Exp Biol 32:668–671

Dias MC, Pinto G, Santos C (2011) Acclimatization of micropropagated plantlets induces an antioxidative burst: a case study with *Ulmus minor* Mill. Photosynthetica 49:259–266

Distabanjong K, Geneve RL (1997) Multiple shoot formation from cotyledonary node segments of Eastern redbud. Plant Cell Tiss Org Cult 47:247–254

Douglas DC, McNamara J (2000) Shoot regeneration from seedling explants of *Acacia mangium* Willd. In Vitro Cell Dev Biol-Plant 36:412–415

Driver JA, Kuniyuki AH (1984) *In vitro* propagation of Paradox walnut rootstock. HortSci 19:507–509

Duhoky MMS, Rasheed KA (2010) Effect of different concentrations of Kinetin and NAA on micropropagation of *Gardenia jasminoides*. J Zankoy Sulaimani 13:103–120

Durkovic J (2008) Micropropagation of mature *Cornus mas* 'Macrocarpa.' Trees 22:597–602

Durkovic J, Lux A (2010) Micropropagation with a novel pattern of adventitious rooting in American sweetgum (*Liquidambar styraciflua* L.). Trees 24:491–497

Durkovic J, Misalova A (2008) Micropropagation of temperate noble hardwoods: An overview. Funct Plant Sci Biotechnol 2:1–19

Economou AS, Spanoudaki MJ (1985) *In vitro* propagation of *Gardenia*. HortSci 20:213

Edson JL, Wenney DL, Leege-Bruven A (1994) Micropropagation of Pacific Dogwood. HortSci 29:1355–1356

Elangomathvan R, Prakash S, Kathiravan K, Seshadari S, Igancimuthu S (2003) High frequency *in vitro* propagation of kidney tea plant. Plant Cell Tiss Org Cult 72:83–86

Elkonin LA, Pakhomova NV (2000) Influence of nitrogen and phosphorus on induction embryogenic callus of sorghum. Plant Cell Tiss Org Cult 61:115–123

Engelmann F (1997) In vitro conservation methods. In: Ford-Lloyd BV, Newbury JH, Callow JA (eds) Biotechnology and plant genetic resources: conservation and Use. CABI, Wallingford, p 119–162

Epstein E, Bloom AJ (2005) Mineral nutrition of plants: principles and perspectives. Sinauer Associates, 2nd edn. Sunderland

Epstein E, Ludwig-Muller J (1993) Indole-3-butyric acid in plants: occurrence, synthesis, metabolism and transport. Physiol Plant 88:382–389

Eswara JP, Stuchbury T, Allan EJ (1998) A standard procedure for the micropropagation of the neem tree (*Azadirachta indica* A. Juss). Plant Cell Rep 17:215–219

Faisal M, Ahmad N, Anis M (2006) *In vitro* plant regeneration from alginate-encapsulated microcuttings of *Rauvolfia tetraphylla* L. World J Agri Sci 1:1–6

Faisal M, Anis M (2003) Rapid mass propagation of *Tylophora indica* Merill. via leaf callus culture. Plant Cell Tiss Org Cult 75:125–129

Faisal M, Anis M (2007) Regeneration of plants from alginate encapsulated shoots of *Tylophora indica* (Burm. f.) Merrill, an endangered medicinal plant. J Hortic Sci Biotechnol 82:351–354

Faisal M, Anis M (2009) Changes in photosynthetic activity, pigment composition, electrolyte leakage, lipid peroxidation and antioxidant enzymes during *ex vitro* establishment of micropropagated *Rauvolfia tetraphylla* plantlets. Plant Cell Tiss. Org. Cult. 99:125–132

Faisal M, Anis M (2010) Effect of light irradiations on photosynthetic machinery and antioxidative enzymes during *ex vitro* acclimatization of *Tylophora indica* plantlets. J Plant Interact 5:21–27

Faisal M, Siddique I, Anis M (2006a) *In vitro* rapid regeneration of plantlets from nodal explants of *Mucuna pruriens*- a valuable medicinal plant. Ann Appl Biol 148:1–6

Faisal M, Siddique I, Anis M (2006b) An efficient plant regeneration system for *Mucuna pruriens* L. (DC) using cotyledonary node explants. In Vitro Cell Dev Biol Plant 42:59–64

Famiani F, Ferradini N, Staffolani P, Standardi A (1994) Effect of leaf excision time, age, BA concentration and dark treatments on *in vitro* shoot regeneration of M26 apple rootstock. J Hort Sci 69:679–685

Farhatullah AZ, Abbas SJ (2007) *In vitro* effects of gibberellic acid on morphogenesis of potato explants. Int J Agri Biol 9:181–182

Ferreira FJ, Kieber JJ (2005) Cytokinin signalling. Curr Opin Plant Biol 8:518–525

Fiedler H (1936) Entwicklungs und reizphysiologische Untersuchungen an Kulturen isolierter Wurzelspitzen. Zeitschrift fur Botanik 30:385–436

Frabetti M, Gutierrez-Pesce P, Mendoza-de Gyves E, Rugini E (2009) Micropropagation of *Teucrium fructicans* L., an ornamental and medicinal plant. In Vitro Cell Dev Biol-Plant 45:129–134

Gaba VP (2005) Plant growth regulators in plant tissue culture and development. In: Trigano RN;Gray DJ. (eds) Plant development and biotechnology. CRC Press, Boca Raton, p 87–99

Gahan PB, George EF (2008) Adventitious regeneration. In: George EF, Hall MA, De Klerk GJ (eds) Plant propagation by tissue culture, 3rd edn. Springer, Dordrecht, p 355–401

Gamborg OL, Miller RA, Ojima K (1968) Nutrients requirement of suspension culture of soybean root cells. Exp Cell Res 50:151–158

Gaspar T, Kevers C, Hausman JF, Ripetto V (1994) Peroxidase activity and endogenous free auxin during adventitious root formation. In: Lumsden PJ, Nicholas JR and Davies WJ (eds.) Physiology, growth and Development of plants in culture. Kluwer Academic Publishers, The Hague, p 289–298

Gautheret RJ (1934) Culture du tissues cambial. C. R. Hebd. Seances Acad Sci 198:2195–2196

Gautheret RJ (1937) La culture des tissus vegetaux. Son etat actuel, comparaison avec la culture des tissus animaux. Hermann & Cie, Paris

Gautheret RJ (1939) Sur la possibilite de realiser a culture indefinite des tissues de tubercules de carotte. C R Hebd Seances Acad Sci 208:118–120

Gautheret RJ (1983) Plant tissue culture: a history. Bot mag Tokyo 96:393–410

Gautheret RJ (1985) History of plant tissue and cell culture: A personal account. In: Vasil IK, ed. Cell culture and somatic cell genetics of plants, Vol. 2. New York: Academic Press, p 1–59

Gauthret RJ (1940) Nouvelles recherche sur le bouregeonnement du tissue cambial d' *Ulmus campestris* cultive *in vitro*. C.R. Acad Sci 210:744–746

Geiger-Huber M, Burlet E (1936) Ueber den hormonalen einflusz der a Indolylessigsaure auf das Wachstums isolierter Wurzeln in keimfreier Organkultur. Jahrbuch fur Wissenschaftliche Botanik 84:233–253

George EF (1993) Plant propagation by tissue culture, part I: the technology. Exegetics, Edington. p 1–574

George EF, Debergh PC (2008) Micropropagation: uses and methods. plant propagation by tissue culture 3rd edn, Vol.1. In: George EF, Hall MA, De Klerk G-J (edn) The background. Published by Springer, Dordrecht

George EF, Sherrington PD (1984) In: Plant propagation by Tissue culture. Exegetics Ltd., London, p 39–71

Georges D, Chemienx JC, Ochatt SD (1993) Plant regeneration from aged callus of woody ornamental species *Lonicera japonica* cv. Hell's prolific. Plant Cell Rep 13:91–94

Germana M, Micheli M, Chiancone B, Macaluso L, Standardi A (2011) Organogenesis and encapsulation of *in vitro*-derived propagules of *Carrizo citrange*, *Citrus sinensis* (L.); *Poncirius trifoliata* (L.). Plant Cell Tiss Org Cult 78:1–9

Gharyal PK, Maheshwari SC (1990) Differentiation of explants from mature leguminous trees. Plant Cell Rep 8:550–553

Giri CC, Shyankumar B, Anjaneyulu C (2004) Progress in tissue culture, genetic transformation and applications of biotechnology to trees: An Overview. Trees 18:115–135

Girijashankar V (2011) Micropropagation of multipurpose medicinal tree, *Acacia auriculiformis*. J Med Plant Res 5:462–466

Gitonga LN, Gichuki ST, Ngamau K, Muigai AWT, Kahangi EM, Wasilwa LA, Wepukhulu S, Njogu N (2010) Effect of explant type, source and genotype on *in vitro* shoot regeneration in Macadamia (*Macadamia* spp.) J Agri Biotechnol Sustain Dev 2:129–135

Gokhale M, Bansal YK (2009) Direct *in vitro* regeneration of a medicinal tree *Oroxylum indicum* (L.) Vent. through tissue culture. Afri J Biotechnol 8:3777–3781

Gonzalez-Benito ME, Perez C, Viviani AB (1997) Cryopreservation of nodal explants of an endangered plant species (*Centaurium rigualii* Esteve) using the encapsulation dehydration method. Biodiversity Conserv 6:583–590

Gonzalez-Rodriguez JA, Ramirez-Garduza F, Manuel L, Robert-Aileen O'Connor-Sanchez, Pena-Ramirez YJ (2010) Adventitious shoot induction from adult tissues of the tropical timber tree yellow Ipé primavera (*Tabebuia donnell-smithii rose* [bignoniaceae]). In Vitro Cell Dev Biol-Plant 46:411–421

Goto S, Thakur RC, Ishii K (1998) Determination of genetic stability in long-term micropropagated shoots of *Pinus thunbergii* Parl. using RAPD markers. Plant Cell Rep 18:193–197

Gratao PL, Polle A, Lea PJ, Azevedo RA (2005) Making the life of heavy metal stressed plants a little easier. Funct Plant Biol 32:481–494

Grout BWW (1988) Photosynthesis of regenerated plantlets *in vitro*, and the stress of transplanting. Acta Hort 230:129–135

Guan QZ, Guo YH, Sui XL, Li W, Zhang ZX (2008) Changes in photosynthetic capacity and oxidant enzymatic systems in micropropagated *Zingiber officinale* plantlets during their acclimation. Photosynthetica 46:193–201

Gubis J, Lajchova Z, Frago J, Jurekova Z (2003) Effect of explant type on shoot regeneration in tomato (*Lycopersicon esculentum* Mill.) *in vitro*. Czech J Plant Breed 39:9–14

Guha S, Maheshwari SC (1966) Cell division and differentiation of embryos in the pollen grains of *Datura in vitro*. Nature 212:97–98

Gupta S, Madal BB (2003) *In vitro* methods for PGR conservation: principles and prospects. In:Chaudhury R, Pandey R, Malik SK, Bhag Mal, editors, *In vitro* conservation and cryopreservation of tropical fruit species. New Delhi: IPGRI office for South Asia and NBPGR 71–80

Gupta PK, Nadgir AI, Mascarenhas AF, Jaganathan V (1980) Tissue culture of forest trees—Clonal multiplication of *Tectona grandis* (teak) by tissue culture. Plant Sci Lett 17:259–268

Gururaj HB, Giridhar P, Ravishankar GA (2007) Micropropagation of *Tinospora cordifolia* (Willd.) Miers ex Hook. F & Thomas—a multipurpose medicinal plant. Curr Sci 92:23–26

Haagen-Smit AJ (1951) The history and nature of plant growth hormones. In: Skoog, F, ed. *Plant growth substances*. Madison: University of Wisconsin Press 3–19

Haberlandt G (1902) Kulturversuche mit isolierten Pflanzenzellen. Sitzungsber K Preuss Akad Wiss Wien. Math Naturwiss 111:69–92.

Haissig BE (1974) Influence of auxins and auxin synergists on adventitious root primodium initiation and development. Bio Tech 5:52–59

Hammatt N, Grant NJ (1998) Shoot regeneration from leaves of *Prunus serotina* Ehrh. (black cherry) and *P. avium* L. (wild cherry). Plant Cell Rep 17:526–530

Haramaki C (1971) Tissue culture of *Gloxinia*. Comb. Proc Int Plant Prop Soc 21:442–448

Haramaki C, Murashige T (1972) *In vitro* culture of *Gloxinia*. Hort Sci 7:339

Harbage JF, Stimart DP, Auer C (1998) pH affects 1 H-indole- 3-butyric acid uptake but not metabolism during the initiation phase of adventitious root induction in apple microcuttings. J Am Soc Hortic Sci 123:6–10

Harisaranraj R, Suresh K, Saravanbabu S (2009) Rapid clonal propagation of *Rauvolfia tetraphylla* L. Acad J Plant Sci 2:195–198

Hausen A, Pal M (2003) Effect of serial bud grafting and etiolation on rejuvenation and rooting cuttings of mature trees of *Tectona grandis* Linn. f. Silvae Genet 52:84–88

Hazarika BN (2003) Acclimatization of tissue cultured plants. Curr Sci 85:1705–1712

Hazarika BN (2006) Morpho-physiological disorders in *in vitro* culture of plants. Sci Hortic 108:105–120

Hazarika BN, Parthasarathy VA, Nagaraju V, Bhowmik G (2000) Sucrose induced biochemical changes in *in vitro* microshoots of Citrus species. Indian J Hortic 57:27–31

Heller R (1953) Researches on the mineral nutrition of plant tissues. Ann Sci Nat Bot Biol Veg 11th Ser 14:1–223

Heyl A, Schmulling T (2003) Cytokinin signal perception and transduction. Curr Opin Plant Biol 6:480–488

Hildebrandt AC, Riker AJ, Duggar BM (1946) The influence of the composition of the medium on growth *in vitro* of excised tobacco and sunflower tissue cultures. Am J Bot 33:591–597

Hiregoudar LV, Kumar HGA, Murthy HN (2005) *In vitro* culture of *Feronia limonia* (L.) Swingle from hypocotyls and intermodal explants. Biol Plant 49:41–45

Hisajima S (1982) Multiple shoot formation from almond seeds and an excised single shoot. Agric Biol Chem 46:1091–1093

Hisajima S, Church L (1981) Multiple shoot formation from soybean embryo. Plant Physiol 67:28

Hossain SN, Munshi MK, Islam MR, Hakim L, Hossain M (2003) *In vitro* propagation of Plum (*Zyziphus jujuba* Lam.). Plant Cell Tiss Org Cult 13:81–84

Hossain SN, Rahman S, Joydhar A, Islam S, Hossain M (2001) *In vitro* propagation of *Acacia catechu* Willd. Plant Tiss Cult 11:25–29

Howell SH, Lall S, Che P (2003) Cytokinins and shoot development. Trends Plant Sci 8:453–459

Howell SH, Lall S, Che P (2003) Cytokinins and shoot development. Trends Plant Sci 8:453–459

Huang FH, Jameel M, Al-Khayri, Gbur EE (1994) Micropropagation of *Acacia mearnsii*. In Vitro Cell Dev Biol-Plant 30:70–74

Huang WJ, Ning GG, Liu GF, Bao MZ (2009) Determination of genetic stability of long-term micropropagated plantlets of *Platanus acerifolia* using ISSR markers Biol Plant 53:159–163

Huetteman CA, Preece JE (1993) Thidiazuron: a potent cytokinin for woody plant tissue culture. Plant Cell Tiss Org Cult 33:105–119

Husain MK, Anis M, Shahzad A (2007) *In vitro* propagation of Indian Kino (*Pterocarpus marsupium* Roxb.) using thidiazuron. In Vitro Cell Dev Biol Plant 43:59–64

Husain MK, Anis M, Shahzad A (2010) Somatic embryogenesis and plant regeneration in *Pterocarpus marsupium* Roxb. Trees 24:781–787

Hussain TM, Chandrasekhar T, Gopal GR (2008) Micropropagation of *Sterculia urens* Roxb., an endangered tree species from intact seedlings. Afri J Biotechnol 7:095–101

Hussein S, Ibrahim R, Kiong ALP, Fadzillah NE and Daud SK (2005) Multiple shoots formation of an important medicinal plant, *Eurycoma longifolia* Jack. Plant Biotechnol 22:349–351

Hyndman SE, Hasegawa PM, Bressan DA (1982) The role of sucrose and nitrogen in adventitious root formation on cultured rose shoots. Plant Cell Tiss Org Cult 1:229–238

Iapichino G, Airo M (2008) Micropropagation of *Metrosideros excelsa*. In Vitro Cell Dev Biol Plant 44:330–337

Isabel N, Tremblay L, Michaud M, Tremblay FM, Bousquet J (1993) RAPDs as an aid to evaluate the genetic integrity of somatic embryogenesis-derived populations of Picea mariana (Mill.) B.S.P. Theor Appl Genet 86:81–87

Islam R, Hossain M, Joarder OI, Karim MR (1993) Adventitious shoot formation on excised leaf explants of *in vitro* grown seedlings of *Aegle marmelos* Corr. J Hortic Sci 68:95–498

Isutsa DK, Pritts MP, Mudge KW (1994) Rapid propagation of blueberry plants using *ex vitro* rooting and controlled acclimatization of micropropagules. Hort Sci 29:1124–1126.

Jablonski JR, Skoog F (1954) Cell enlargement and cell division in excised tobacco pith tissue. Physiol Plant 7:16–24

Jahan AA, Anis M (2009) *In vitro* rapid multiplication and propagation of *Cardiospermum halicacabum* L. through axillary bud culture. Acta Physiol Plant 31:133–138

Jain N, Babbar SB (2000) Recurrent production of plants of black plum. *Syzygium cumini* (L.) Skeels, a myrtaceous fruit tree, from *in vitro* cultured seedling explants. Plant Cell Rep 19:519–524

Jakola L, Tolvanen A, Laine K, Hohtola A (2001) Effect of N^6-isopentenyladenine concentration on growth initiation *in vitro* and rooting of berberry and lingonberry microshoots. Plant Cell Tiss Org Cult 66:73–77

Jeong JH, Murthy HN, Paek KY (2001) High frequency adventitious shoot induction and plant regeneration from leaves of statice. Plant Cell Tiss Org Cult 65:123–128

Jha S, Jha TB (1989) Micropropagation of *Cephalis ipecacuanha* Rich. Plant Cell Rep 8:437–439

Jimenez A, Hernandez JA, Delrio LA, Sevilla F (1997) Evidence for the presence of the ascorbate–glutathione cycle in mitochondria and peroxisomes of pea leaves. Plant Physiol 114:275–284

Jo EA, Tewari RK, Hahn EJ, Paek KY (2009) *In vitro* sucrose concentration affects growth and acclimatization of *Alocasia amazonica* plantlets. Plant Cell Tiss Org Cult 96:307–315

Jones MPA, Yi Z, Murch SJ, Saxena PK (2007) Thidiazuron-induced regeneration of *Echinacea purpurea* L.: micropropagation in solid and liquid culture systems. Plant Cell Rep 26:13–19

Joshi P, Dhawan V (2007) Assessment of genetic fidelity of micropropagated *Swerita chiraytia* plantlets by ISSR marker assay. Biol Plant 51:22–26

Kadlecek P, Ticha I, Capkova V, Schafer C (1998) Acclimatization of micropropagated tobacco plantlets. In: Garab G (ed) Photosynthesis: mechanisms and effects vol. V. Kluwer Academic Publishers, Dordrecht, p 3853–3856

Kaneda Y, Tabei Y, Nishimura S, Harada K, Akihama T and Kitamura K (1997) Combination of thidiazuron and basal media with low salt concentrations increases the frequency of shoot organogenesis in soybeans (*Glycine max* (L.) Merr.). Plant Cell Rep 17:8–12

Kanthrajah AS, Rechard GD, Dodd WA (1992) roots as source of explants for successful micropropagation of Cacambola (*Averrhoa carambola*). Scientia Horti 51:169–177

Karsai I, Bedo Z, Hayes PM (1994) Effect of induction medium pH and maltose concentration on *in vitro* androgenesis of hexaploid winter triticale and wheat. Plant Cell Tiss Org Cult 39:49–53

Kartha KK (1981) Meristem culture and cryopreservation- Methods and applications. In: Thorpe TA (ed.) Plant Tissue Culture: Methods and Applications in Agriculture. Academic Press, New York, p 181–211

Karuppanapandian T, Moon JC, Kim C, Manoharan K, Kim W (2011) Reactive oxygen species in plants: their generation, signal transduction and scavenging mechanisms. Aust J Crop Sci 5:709–725

Kaveriappa KM, Phillips LM, Trigiano RN (1997) Micropropagation of flowering dogwood (*Cornus florida*) from seedlings. Plant Cell Rep 16:485–489

Keil M, Griffin RA (1994) Use of random amplified polymorphic DNA (RAPD) markers in the discrimination and verification of genotypes in *Eucalyptus*. Theor Appl Genet 89:442–450

Kenneth E, Pallet KE, Young J (2000) Carotenoids. In: Ruth GA, Hess JL (eds) Antioxidants in higher plants. CRC Press, USA, p 60–81

Khan A, Chauhan YS, Roberts LW (1986) *In vitro* studies on xylogenesis in *Citrus* fruit vesicles. 11. Effect of pH of the nutrient medium on the induction of cytodifferentiation. Plant Sci 46:213–216

Khan MI, Ahmad N, Anis M (2011) The role of cytokinins on *in vitro* shoot production in *Salix tetrasperma* Roxb.: a tree of ecological importance. Trees 25:577–584

Kilb B, Wietoska H, Godde D (1996) Changes in the expression of photosynthetic genes precede loss of photosynthetic activities and chlorophyll when glucose is supplied to mature spinach leaves. Plant Sci 115:225–235

Kim MK, Sommer HE, Bongorten BC, Merkle SA (1997) High frequency induction of adventitious shoots from hypocotyls segments of *Liquidambar styraciflua* L. by thidiazuron. Plant Cell Rep 16:536–540

Kitto SK, Janick J (1982) Polyox as an artificial seed coat for a sexual embryos. Hort Sci 17:488

Knop W (1865) Quantitative Untersuchungen uber den Ernahrungsprozess der Pflanzen. Landwirtsch Vers Stn 7:93–107

Koroch AR, Jr. Juliani HR, Juliani HR, Trippi VS (1997) Micropropagation and acclimatization of *Hedeoma multiflorum*. Plant Cell Tiss Org Cult 48:213–217

Kotsias D, Roussos PA (2001) An investigation on the effect of different plant growth regulating compounds in *in vitro* shoot tip and node culture of lemon seedlings. Sci Hortic 89:115–128

Kumar N, Vijay Anand KG, Reddy MP (2010) Shoot regeneration from cotyledonary leaf explants of *Jatropha curcas*: a biodiesel plant. Acta Physiol Plant 32:917–924

Kumar R, Sharma K, Agrawal V (2005) *In vitro* clonal propagation of *Holarrhena antidysenterica* (L.) Wall. through nodal explants from mature trees. In Vitro Cell Dev Biol Plant 41:137–144

Kumar S, Mangal M, Dhawan AK, Singh N (2011) Assessment of genetic fidelity of micropropagated plants of *Simmondsia chinensis* (Link.) Schneider using RAPD and ISSR markers. Acta Physiol Plant 33:2541–2545

Kunze I (1994) Influence of the genotype on growth of Norway Spruce (*Picea abies* L.) in *in vitro* meristem culture. Silvae Genet 43:36–41

Kwa SH, Wee YC, Lim TM, Kumar PP (1995) Establishment and physiological analyses of photoautotrophic callus cultures of the fern *Platycerium coronarium* (Koenig) Desv under CO_2 enrichment. J Exp Bot 46:1535–1542

Lakshmisita G, Chattopadhyay S, Tejavati DH (1986) Plant regeneration from shoot callus of rosewood (*Dalbergia latifolia* Roxb.). Plant Cell Rep 5:266–268

Lakshmisita G, Sreenatha KS, Sujata S (1992) Plantlet production from shoot tip cultures of red sandalwood (*Pterocarpus santalinus* L.). Curr. Sci. 62:532–535

Lal N, Ahuja PS, Kukreja AK, Pandey B (1988) Clonal propagation of *Picrorhiza kurroa* Royale ex. Benth. by shoot tip culture. Plant Cell Rep 7:202–205

Lameira OA, Pinto JEBP (2006) *In vitro* propagation of *Cordia verbenaceae* L. (Boraginaceae) Rev. Bras. Plant Med Botucatu 8:102–104

Larkin P, Scowcroft WR (1981) Somaclonal variation, a novel source of variability from cell cultures for plant improvement. Theor Appl Genet 60:197–406

Lavanya M, Venkateshwarlu B, Devi BP (2009) Acclimatization of neem microshoots adaptable to semi-sterile conditions. Indian J Biotechnol 8:218–222

Leblay C, Chevreau E, Raboin LM (1991) Adventitious shoot regeneration from leaves of several pear cultivars (*Pyrus communis* L.). Plant Cell Tiss Org Cult 25:99–105

Leljak-Levanic D, Bauer N, Mihaljevic S, Jelaska S (2004) Somatic embryogenesis in pumpkin (*Cucurbita pepo* L.): control of somatic embryo development by nitrogen compounds. J Plant Physiol 161:229–236

Leroy XJ, Leon K, Charles G, Branchard M (2000) Cauliflower somatic embryogenesis and analysis of regenerant stability by ISSRs. Plant Cell Rep 19:1102–1107

Leyser HMO, Pickett FB, Dharmsiri S, Estelle E (1996) Mutation in AXR3 gene of *Arabidopsis* result in altered auxin response including ectopic expression from the SAUR-AC1 promoter. Plant J 10:403–413

Li G, Qulros CF (2001) Sequence-related amplified polymorphism (SRAP), a new marker system based on a simple PCR reaction: its application to mapping and gene tagging in Brassica. Theor Appl Genet 103:455–461

Li Y, Shi XY, Strabala TJ, Hagen G, Guilfoyle TJ (1994) Transgenic tobacco plants that overproduce cytokinins show increased tolerance to exogenous auxin and auxin transport inhibitors. Plant Sci 100:9–14

Libbenga KR, Mennes AM (1995) Hormone binding and signal transduction. In: Davies PJ (ed.) Plant Hormones: Physiology, biochemistry and molecular biology. 2nd edn., Kluwer Publishers, Dordrecht, p 272–297

Lloyd G, McCown B (1980) Commercially feasible micropropagation of mountain laurel, *Kalmia latifolia*, by use of shoot tip culture. Int Plant Prop Soc Proc 30:421–427

Loc HN, Due TD, Kwon HT, Yang SM (2005) Micropropagation of zedoary (*Curcuma zedoaria* Roscoe): a valuable medicinal plant. Plant Cell Tiss Org Cult 81:119–122

Lombardi SP, Passos IRS, Nogueira MCS, Appezato-da-Gloria B (2007) *In vitro* shoot regeneration from roots and leaf discs of *Passiflora cincinnata* Mast. Braz Arch Biol Biotechnol 50:239–247

Loo SW (1945a) Cultivation of excised stem tips of *Asparagus in vitro*. Am J Bot 32:13–17

Loo SW (1945b) Cultivation of excised stem tips of intact plants under sterile conditions. Ph. D. Thesis. Pasadena: California Institute of Technology.

Loo SW (1946a) Further experiments on the culture of excised asparagus tips *in vitro*. Am J Bot 33:156–159

Loo SW (1946b) Cultivation of excised stem tips of dodder *in vitro*. Am J Bot 33:295–300

Lu CY (1993) The use of thidiazuron in tissue culture. In Vitro Cell Dev Biol-Plant 29:92–96

Ludwig-Muller J (2000) Indole-3-butyric acid in plant growth and development. Plant Growth Regul 32:219–230

Lynch PT (1999) Tissue culture techniques in *in vitro* plant conservation. In: Benson EE (ed) Plant conservation biotechnology. Taylor & Francis, London

Lyyra S, Lima A, Merkle S (2006) *In vitro* regeneration of *Salix nigra* from adventitious shoots. Tree Physiol 26:969–975

Magyar-Tabori K, Dobranszki J, Teixeira da Silva JA, Bulley SM, Hudak I (2010) The role of cytokinins in shoot organogenesis in apple. Plant Cell Tiss Org Cult 101:251–267

Malabadi RB, Staden JV (2005) Storability and germination of sodium alginate encapsulated somatic embryos derived from the vegetative shoot apices of mature *Pinus patula* trees. Plant Cell Tiss Org Cult 82:259–265

Malik KA, Saxena PK (1992a) Thidiazuron induces high-frequency shoot regeneration in intact seedlings of pea (*Pisum sativum*), chickpea (*Cicer arietinum*) and lentil (*Lens culinaris*). Aust J Plant Physiol 19:731–740

Malik KA, Saxena PK (1992b) *In vitro* regeneration of plants: a novel approach. Naturwissenschaften 79:136–137

Mallikarjuna K, Rajendurdu G (2007) High frequency *in vitro* propagation of *Holarrhena antidysentrica* from nodal buds of mature tree. Biol Plant 51:525–529

Mante SR Scorza, Cordts JM (1989) Plant regeneration from cotyledons of *Prunus persica*, *Prunus domestica*, and *Prunus cerasus*. Plant Cell Tiss Org Cult 19:1–11

Martin G, Geetha SP, Raja SS, Raghu AV, Balachandran I, Ravindran PN (2006) An efficient micropropagation system for *Celastrus paniculatus* Willd.: a vulnerable medicinal plant. J Fores Res 11:461–46

Martin SM, Rose D (1976) Growth of plant cell (*Ipomea*) suspension cultures at controlled pH levels. Can J Bot 54:1264–1270

Martins M, Sarmento D, Oliveira MM (2004) Genetic stability of micropropagated almond plantlets, as assessed by RAPD and ISSR markers. Plant Cell Rep 23:492–496

McCoy TJ, Smith LY (1986) Interspecific hybridization of perennial *Medicago* species using ovule-embryo culture. Tag 71:772–783

McGaw BA, Burch LR (1995) Cytokinin biosynthesis and metabolism. In: Davies PJ (ed) Plant hormones, physiology, biochemistry and molecular biology. Kluwer, Dordrecht, p 98–117

Mehta UJ, Krishnamurthy KV, Hazra S (2000) Regeneration of plants via adventitious bud formation from zygotic embryo axis of tamarind (*Tamarindus indica* L.). Curr Sci 78:1231–1234

Meijer EGM, Brown DCW (1987) Role of exogenous reduced nitrogen and sucrose in rapid high frequency somatic embryogenesis in *Medicago sativa*. Plant Cell Tiss Org Cult 10:11–19

Merkle SA, Nairn CJ (2005) Hardwood tree biotechnology. In Vitro Cell Dev Biol-Plant 41:602–619

Micheli M, Hafiz IA, Standardi A (2007) Encapsulation of in vitro derived explants of olive (Olea europaea L. cv. Moraiolo):II. Effects of storage on capsule and derived shoot performance. Sci Hortic 113:286–292

Miller CO, Skoog F, Von Saltza MH, Strong FM (1955) Kinetin, a cell division factor from deoxyribonucleic acid. J Am Chem Soc 77:1392

Mills D, Wenkart S, Benzioni A (1997) Micropropagation of Simmondsia chinensis (jojoba). In: Bajaj YPS (ed) Biotechnology in Agriculture and Forestry

Mishra J, Singh M, Palni LMS, Nandi SK (2011) Assessment of genetic fidelity of encapsulated microshoots of Picrorhiza kurrooa. Plant Cell Tiss Org Cult 104:181–186

Mishra P, Datta SK (2001) Acclimatization of Asiatic hybrid lilies under stress conditions after propagation through tissue culture. Curr Sci 81:1530–1533

Mittler R (2002) Oxidative stress, antioxidants and stress tolerance. Trends Plant Sci 7:405–410

Mok MC, Mok DW, Amstrong DJ, Shudo K, Isogai Y, Okamanto T (2005) Cytokinin activity of N-phenyl-N0–1,2,3-thiadiazol-5-urea (thidiazuron). Phytochem 21:1509–1511

Molle F, Dupis JM, Ducos JP, Anselm A, Corulus-Savidan I, Petiard V, Freyssinel G (1993) Carrot somatic embryogenesis and its application to synthetic seeds. In: (Redenbaugh K ed) Synseeds, Application of synthetic seeds to crop improvement. CRC Press Boca Raton FL, USA, p 257–287

Moller IM, Jensen PE, Hansson A (2007) Oxidative modifications to cellular components in plants. Annu Rev Plant Biol 58:459–481

Morel G (1964) Tissue culture—a new means of clonal propagation of orchids. Am Orchid Soc Bul 33:473–478

Morimoto M, Nakamura K, Sano H (2006) Regeneration and genetic engineering of a tropical tree, Azadirachta excelsa. Plant Biotechnol 23:123–127

Morton L, Browse J (1991) Facile transformation of Arabidopsis. Plant Cell Rep 10:235–239

Moshkov IE, Novikova GV, Hall MA and George EF (2008) Plant growth regulators III: gibberellins, ethylene, abscisic acid, their analogues and inhibitors; miscellaneous compounds. In: George EF, Hall MA, De Klerk GJ (eds) Plant propagation by tissue culture. Springer, The Netherlands, p 227–282

Murashige T (1974) Plant Propagation through tissue culture. Ann Rev Plant Physiol 25:135–166

Murashige T, Skoog F (1962) A revised medium for rapid growth and bioassays with tobacco tissue cultures. Physiol Plant 15:473–497

Murthy BNS, Murch SJ, Saxena PK (1988) Thidiazuron: a potent regulator of in vitro plant morphogenesis. In Vitro Cell Dev Biol-Plant 34:267–272

Mylona P, Dolan L (2002) The root meristem. In: McManus MT, Weit BE (eds) Meristematic tissues in plant growth and development. Sheffield Academic, Sheffield, p 279–292

Nagaoka T, Ogihara Y (1997) Applicability of inter-simple sequence repeat polymorphism in wheat for use as DNA markers in comparision to RFLP and RAPD markers. Theor Appl Genet 94:597–602

Naik D, Vartak V, Bhargava S (2003) Provenance- and subculture-dependent variation during micropropagation of Gmelina arborea. Plant Cell Tiss Org Cult 73:189–195

Naik PM, Manohar SH, Praveen N, Murthy HN (2010) Effects of sucrose and pH levels on in vitro shoot regeneration from leaf explants of Bacopa monnieri and accumulation of bacoside A in regenerated shoots. Plant Cell Tiss Org Cult 100:235–239

Naik SK, Chand PK (2006) Nutrient-alginate encapsulation of in vitro nodal segments of pomegranate (Punica granatum L.) for germplasm distribution and exchange. Sci Hortic 108:247–252

Nair LG, Seeni S (2001) Rapid in vitro multiplication and restoration of Celastrus paniculatus (Celastraceae), a medicinal woody climber. Indian J Exp Biol 39:697–704

Nandwani D, Mathur N, Ramawat KG (1995) In vitro shoot multiplication from cotyledonary node explants of Tecomella undulata. Gartenbauwisseschaft 60:65–68

Nas MN, Read PE (2004) Improved rooting and acclimatization of micropropagated hazelnut shoots. Hort Sci 39:1688–1690

Nayak P, Behera PR, Manikkannan T (2007) High frequency plantlet regeneration from cotyledonary node cultures of Aegle marmelos (L.) Corr. In Vitro Cell Dev Biol-Plant 43:231–236

Naz S, Ali A, Siddique FA, Iqbal J (2008) Somatic embryogenesis from immature cotyledons and leaf calli of chick pea (Cicer arietinum L.). Pak J Bot 40:523–531

Nedelcheva S (1986) Effect of inorganic components of the nutrient medium on in vitro propagation of pears. Genet Sel 19:404–406

Nef-Campa C, Chaintreuil-Dongmo C, Dreyfus BL (1996) Regeneration of the tropical legume Aeschynomene sensitiva Sw. from root explants. Plant Cell Tiss Org Cult 44:149–154

Niederwieser JG, Van Staden J (1990) the relationship between genotype, tissue age and endogenous cytokinin levels on adventitious bud formation on leaves of Lachenalia. Plant Cell Tiss Org Cult 22:223–228

Nitsch JP, Nitsch C (1956) Auxin-dependent growth of excised Helianthus tissues. Am J Bot 43:839–851

Nobecourt P (1939) Sur la perennite et l'augmentation de volume des cultures de tissues vegetaux. C. R. Seances Soc Biol Ses Fil 130:1270–1271

Noctor G, Foyer CH (1998) Ascorbate and glutathione: keeping active oxygen under control. Annu Rev Plant Physiol Plant Mol Biol 49:249–279

Noctor G, Gomez L, Vanacker H, Foyer CH (2002) Interactions between biosynthesis, compartmentation and transport in the control of gluthatione homeostasis and signalling. J Exp Bot 53:1283–1304

Nunes EDC, de Castilho CV, Moreno FN, Viana AM (2002) *In vitro* culture of *Cedrela fissilis* Vellozo (Meliaceae). Plant Cell Tiss Org Cult 70:259–268

Nyende A, Schittenhelm S, Mix-Wagner G, Greef JM (2003) Production, storability, and regeneration of shoot tips of potato (*Solanum tuberosum* L.) encapsulated in calcium alginate hollow beads. In Vitro Cell Dev Biol-Plant 39:540–544

Ogunsola KE, Ilori CO (2007) *In vitro* propagation of miracle berry (*Synsepalum dulcificum* Daniel) through embryo and nodal cultures. Afr J Biotechnol 7:244–248

Onay A (2000) Somatic embryogenesis in cultured kernels of pistachio, *Pistacia vera* L. cv. Siirt. Proceedings of the 2nd Balkan Botanical Congress, Istanbul, May 14–16, 2000. Plants of the Balkan Peninsula: Into the Next Millennium 11:109–115

Ozias-Akins P, Vasil IK (1985) Nutrition of plant tissue cultures. In: Vasil IK (ed) Cell culture and somatic cell genetics of plants. Cell growth, nutrition, cytodifferentiation and cryopreservation, Vol 2. Academic Press, New York, p 129–147

Ozudogru EA, Kaya E, Kirdok E, Issever-Ozturk S (2011) *In vitro* propagation from young and mature explants of thyme (*Thymus vulgaris* and *T. longicaulis*) resulting in genetically stable shoots. In Vitro Cell Dev Biol-Plant 47:309–320

Panda BM, Hazra S (2010) *In vitro* regeneration of *Semecarpus anacardium* L. from axenic seedling-derived nodal explants. Trees 24:733–742

Pandey S, Jaiswal VS (2002) Micropropagation of *Terminalia arjuna* Roxb. from coteledonary nodes. Indian J Exp Biol 40:950–953

Pandey S, Singh M, Jaiswal U, Jaiswal VS (2006) Shoot initiation and multiplication from a mature tree of *Terminalia arjuna* Roxb. In Vitro Cell Dev Biol-Plant 42:389–393

Pant M, Bisht P, Gusain MP (2010) De novo Shoot Organogenesis from cultured root explants of *Swertia chirata* Buch.Ham.ex Wall.: An endangered medicinal plant. Nat Sci 8:244–252

Park YS, Barret JD, Bonga JM (1998) Application of somatic embryogenesis in high-value clonal forestry: Deployment, genetic control, and stability of cryopreserved clones. In Vitro Cell Dev Biol-Plant 34:231–239

Parliman BJ, Evans PT, Rupert EA (1982) Tissue culture of single rhizome explants of *Dionaea muscipula* Ellis ex. L., the venus fly trap for rapid asexual propagation. Hort Sci 107:305–310

Parveen S, Shahzad A (2011) A micropropagation protocol for *Cassia angustifolia* Vahl. from root explants. Acta Physiol Plant 33:789–796

Parveen S, Shahzad A, Saema S (2010) *In vitro* plant regeneration system for *Cassia siamea* Lam., a leguminous tree of economic importance. Agrofores Syst 80:109–116

Pasqua G, Manes F, Monacelli B, Natale L, Anselmi S (2002) Effects of the culture medium pH and ion uptake in *in vitro* vegetative organogenesis in thin cell layers of tobacco. Plant Sci 162:947–955

Pattnaik J, Debata BK (1996) Micropropagation of *Hemidesmus indicus* (L.) R. Br. through axillary bud culture. Plant Cell Rep 15:427–430

Pattnaik S, Chand PK (2000) Morphogenic response of the alginate encapsulated axillary buds from *in vitro* shoot cultures of six mulberries. Plant Cell Tiss Org Cult 64:177–85

Pattnaik SK, Chand PK (1997) Rapid clonal propagation of three mulberries, *Morus cathayana* Hemsl., *Morus ihou* Koiz., and *Morus serrata* Roxb., through *in vitro* culture of apical shoot bud and nodal explants from mature trees. Plant Cell Rep 16:503–508

Pattnaik SK, Chand PK, Naik SK, Pradhan C (2000) shoot organogenesis and plant regeneration from hypocotyl-derived cell suspensions of a tree legume, *Dalbergia sissoo* Roxb. In Vitro Cell Dev Biol-Plant 36:407–411

Pedroso C, Oliveira M, Pais SS (1992) Micropropagation and simultaneous rooting of *Actinidia deliciosa* var. 'Hayward'. HortSci 27:443–445

Pena-Ramirez YJ, Juarez-Gomez J, Gomez-Lopez L, Jeronimo-Perez JL, Garcia-Shesena I, Gonzalez-Rodriguez JA, Robert ML (2010) Multiple adventitious shoot formation in Spanish Red Cedar (*Cedrela odorata* L.) cultured *in vitro* using juvenile and mature tissues: an improved micropropagation protocol for a highly-valuable tropical tree species. In Vitro Cell Dev Biol-Plant 46:149–160

Pereira GJG, Molina SMG, Lea PJ, Azevedo RA (2002) Activity of antioxidant enzymes in response to cadmium in *Crotalaria juncea*. Plant Soil 239:123–132

Pereira MJ (2006) Conservation of *Vaccinium cylindraceum* Smith (Ericaceae) by micropropagation using seedling nodal explants. In Vitro Cell Dev Biol Plant 42:65–68

Perveen S, Varshney A, Anis M, Aref IM (2011) Influence of cytokinins, basal media and pH on adventitious shoot regeneration from excised root cultures of *Albizia lebbeck*. J Fores Res 22:47–52

Phillips GC, Collins GB (1979) *In vitro* tissue culture of selected legumes and plant regeneration from callus of red clover. Crop Sci 19:59–64

Phulwaria M, Rai MK, Harish, Gupta AK, Ram K, Shekhawat NS (2012) An improved micropropagation of *Terminalia bellirica* from nodal explants of mature tree. Acta Physiol Plant 34:299–305

Pinto G, Santos C, Neves L, Araújo C (2002) Somatic embryogenesis and plant regeneration in *Eucalyptus globulus* Labill. Plant Cell Rep 21:208–213

Pintos B, Bueno M, Cuenca B, Manzanera J (2008) Synthetic seed production from encapsulated somatic embryos of cork oak (*Quercus suber* L.) and automated growth monitoring. Plant Cell Tiss Org Cult 95:217–225

Poddar K, Vishnoi RK, Kothari SL (1997) Plant regeneration from embryogenic callus of finger millet *Eleusine coracana* (L.) Gaertn. on higher concentrations of NH_4NO_3 as a replacement of NAA in the medium. Plant Sci. 129:101–106

Pospisilova J, Synkova H, Haisel D, Catsky J, Wilhelmova N, Sramek F (1999a) Effect of elevated CO_2

concentration on acclimatization of tobacco plantlets to *ex vitro* conditions. J Exp Bot 50:119–126

Pospisilova J, Ticha I, Kadlecek P, Haisel D, Plzakova S (1999b) Acclimatization of micropropagated plants to *ex vitro* conditions. Biol Plant 42:481–497

Powling A, Hussey G (1981) Stimulation of multiple shoot bud formation in walnut seeds. Hort Sci 17:592

Pradhan C, Kar S, Pattnaik S, Chand PK (1998) Propagation of *Dalbergia sissoo* Roxb. through *in vitro* shoot proliferation from cotyledonary nodes. Plant Cell Rep 18:122–126

Prakash E, Khan PSSV, Rao TJVS, Meru ES (2006) Micropropagation of red sanders (*Pterocarpus santalinus* L.) using mature nodal explants. J Fores Res 11:329–33

Prakash S, Van Staden J (2008) Micropropagation of *Searsia dentata*. In Vitro Cell Dev Biol-Plant 44:338–341

Preece KE, Sutter EG (1991) Acclimatization of micropropagated plants to the greenhouse and field. In: DeBergh PC; Zimmerman RH.(eds) Micropropagation, technology and application. Kluwer Academic, Dordrecht, 71–93

Premkumar A, Mercado JA, Quesada MA (2001) Effects of *in vitro* tissue culture conditions and acclimatization on the contents of Rubisco, leaf soluble proteins, photosynthetic pigments, and C/N ratio. J Plant Physiol 158:835–840

Purkayastha J, Sugla T, Paul A, Solleti S, Sahoo L (2008) Rapid *in vitro* multiplication and plant regeneration from nodal explants of *Andrographis paniculata*, a valuable medicinal plant. In Vitro Cell Dev Biol Plant 44:442–447

Purohit SD, Kukda G (1994) *In vitro* propagation of *Wrightia tinctoria*. Biol Plant 36:519–526

Quoirin M, Lepoivre P (1977) Etude de milieux adaptes aux cultures in vitro de Prunus. Acta Hortic 78:437–442

Quraishi A, Koche V, Sharma P, Mishra SK (2004) *In vitro* clonal propagation of neem (*Azadiracta indica*). Plant Cell Tiss Org Cult 78:281–283

Raghu AV, Geetha SP, Martin G, Balachandran I, Ravindran PN, Mohanan KV (2007) An improved micropropagation protocol for Bael- a vulnerable medicinal tree. Res J Bot 2:186–194

Rai MK, Asthana P, Jaiswal VS, Jaiswal U (2010) Biotechnological advances in guava (*Psidium guajava* L.): recent developments and prospects for further research. Trees 24:1–12

Rai MK, Asthana P, Singh SK, Jaiswal VS, Jaiswal U (2009) The encapsulation technology in fruit plants- A review. Biotechnol Adv 27:671–679

Rai MK, Jaiswal VS, Jaiswal U (2008) Alginate-encapsulation of nodal segments of guava (*Psidium guajava* L.) for germplasms exchange and distribution. J Hortic Sci Biotech 83:569–573

Rajeswari V, Paliwal K (2008) *In vitro* plant regeneration of red sanders (*Pterocarpus santalinus* L. f.) from cotyledonary nodes. Indian J Biotechnol 7:541–546

Rao CD, Goh CJ, Kumar PP (1996) High frequency adventitious shoot regeneration from excised leaves of *Paulownia spp.* cultured *in vitro*. Plant Cell Rep 16:204–209

Rauf S, Rahman H, Khan TM (2004) Effect of Kinetin on multiple shoot induction in cotton (*Gossypium hirsutum* L) cv. NIAB—999. Iranian J Biotechnol 4:279–282

Reddy SP, Rama Gopal G, Laxshmi Sita G (1998) *In vitro* multiplication of *Gymnema sylvestre* R. Br.—An important medicinal plant. Curr Sci 75:843–845

Redenbaugh K, Fuji JA, Slade D (1991) Synthetic seeds technology. In: Vasil IK (ed.) Scale-up and Automation in Plant Propagation. Academic Press, San Diego, p 35–74

Reiter RS, Williams JGK, Feldmann KA, Rafalaski JA, Tingby SV, Scolink PA (1992) Global and local genome mapping in *Arabidopsis thaliana* by using recombinant inbred lines and random amplified polymorphic DNAs. Proc Natl Acad Sci USA 89:1477–1481

Robbins WJ (1922) Cultivation of excised root tips and stem tips under sterile conditions. Bot Gaz 73:376–390

Roberts AW, Haigler CH (1994) Cell expansion and tracheary element differentiation are regulated by extracellular pH in mesophyll cultures of *Zinnia elegans* L. Plant Physiol 105:699–706

Rodriguez R (1982) Stimulation of multiple shoot bud formation in walnut seeds. Hort Sci 17:592

Rolland F, Moore B, Sheen J (2002) Sugar sensing and signaling in plants. Plant Cell 14:185–205

Rout GR, Samantray S, Das P (2000) *In vitro* manipulation and propagation of medicinal plants. Biotechnol Adv 18:91–120

Ruffoni B, Gazzano A, Costantino C (1994) Micropropagazione di Myrtus communis Mill.: Effetto dell'acido indolacetico sulla radicazione. Italus Hortus 1:8–12

Ruiz RM, Rivero HAS, Rudriquez de la OJL, Alcala VMC, Espinosa MAG, Castellanos JS (2005) Micropropagación clonal *in vitro* en *Eucalyptus grandis* y *E. Urophylla*. Ra Ximhai 1:111–130

Sankhla D, Davis TD, Sankhla N (1994) Thidiazuron-induced *in vitro* shoot formation from roots of intact seedlings of *Albizia julibrissin*. Plant Growth Regul 14:267–272

Scarpa MG, Milia M, Satta M (2000) The influence of growth regulators on proliferation and rooting of *in vitro* propagated myrtle. Plant Cell Tiss Org Cult 62:175–179

Schaeffer WI (1990) Terminology associated with cell, tissue and organ culture, molecular biology and molecular genetics. In Vitro Cell Dev Biol-Plant 26:97–101

Schleiden MJ (1838) Beitrage zur Phytogenesis. Arch. Anat. Physiol. Wiss. Med (J Muller) 137–176

Scholten HJ, Pierik RLM (1998) Agar as a gelling agent: chemical and physical analysis. Plant Cell Rep 17:230–235

Schwann T (1839) Mikroscopische Untersuchungen uber die Ubereinstimmung in der Struktur und dem Wachstum des Thiere und Pflanzen. W Engelmann: Leipzig No. 176

Selvakkumar C, Balakrishnan A, Lakshmi BS (2007) Rapid *in vitro* micropropagation of *Alpinia officina-*

rum Hance., an important medicinal plant, through rhizome bud explants. Asi J Plant Sci 6:1251–1255

Serret MD, Trillas MI, Matas J, Araus JL (1997) Development of photoautotrophy and photoinhibition of *Gardenia jasmoides* plantlets during micropropagation. Plant Cell Tiss Org Cult 45:1–16

Sha Valli Khan PS, Prakash E, Rao KR (1997) *In vitro* propagation of an endemic fruit tree *Syzygium alternifolium* (Wight) Walp. Plant Cell Rep 16:325–328

Shahin-uz-zaman M, Ashrafuzzaman H, Shahidul Haque M, Lutfun LN (2008) *In vitro* clonal propagation of the neem tree (*Azadirachta indica* A. Juss). Afri J Biotechnol 7:386–391

Shahzad A, Faisal M, Anis M (2007) Micropropagation through excised root culture of *Clitoria ternatea* and comparison between in vitro regenerated plants and seedlings. Ann Appl Biol 150:341–349

Shan Z, Raemakers K, Tzitzikas EN, Ma Z, Visser RGF (2005) Development of a highly efficient, repetitive system of organogenesis in soybean [*Glycine max* (L.) Merr]. Plant Cell Rep 24:507–512

Sharma AK, Sharma M, Chaturvedi HC (2002) Conservation of phytodiversity of *Azadirachta indica* A. Juss. through *in vitro* strategies, in Role of Plant Tissue Culture in Biodiversity Conservation and Economic Developments, edited by SK Nandi, LMS Palni and A Kumar (Gyanodaya Prakashan, Nainital, India) 513–520

Sharma P, Koche V, Quraishi A, Mishra SK (2005) Somatic embryogenesis in *Buchanania lanzan* spreng. In Vitro Cell Dev Biol-Plant 41:645–647

Shekhawat MS, Shekhawat NS (2011) Micropropagation of *Arnebia hispidissima* (Lehm) DC and production of alkannin from callus and cell suspension culture. Acta Physiol Plant 33:1445–1450

Shepley S, Chen C, Chang JLL (1972) Does gibberellic acid stimulates seed germination via amylase synthesis. Plant Physiol 49:441–442

Shirin F, Rana PK, Mandal AK (2005) *In vitro* clonal propagation of mature *Tectona grandis* through axillary bud proliferation. J Fores Res 10:465–469

Shrivastava S, Banerjee M (2008) *In vitro* clonal propagation of physic nut (*Jatropha curcas* L.): Influence of additives. Int J Integ Biol 3:73–79

Shukla S, Shukla SK, Mishra SK (2009) *In vitro* plant regeneration from seedling explants of *Stereospermum personatum* D.C.: a medicinal tree. Trees 23:409–413

Shyamkumar B, Anjaneyulu C, Giri CC (2003) Multiple shoot induction from cotyledonary node explants of *Terminalia chebula* Retz: a tree of medicinal importance. Biol Plant 47:585–588

Siddique I, Anis M (2007a) *In vitro* shoot multiplication and plantlet regeneration from nodal explants of *Cassia angustifolia* (Vahl.): a medicinal plant. Acta Physiol Plant 29:233–238

Siddique I, Anis M (2007b) High frequency shoot regeneration and plantlet formation in *Cassia angustifolia* (Vahl.) using thidiazuron. Med. Aroma. Plant Sci Biotechnol 1:282–284

Siddique I, Anis M (2008) An improved plant regeneration system and *ex vitro* acclimatization of *Ocimum basilicum* L. Acta Physiol Plant 30:493–499

Siddique I, Anis M (2009b) Morphogenic response of the alginated encapsulated nodal segments and antioxidative enzymes analysis during acclimatization of *Ocimum basilicum* L. J Crop Sci Biotechnol 12:233–238

Siddique I, Anis M, Aref IM (2010) *In Vitro* adventitious shoot regeneration via indirect organogenesis from petiole explants of *Cassia angustifolia* Vahl.—a potential medicinal plant. Appl Biochem Biotechnol 162:2067–2074

Siddique I, Anis M, Jahan AA (2006) Rapid multiplication of *Nyctanthes arbor-tristis* L. through *in vitro* axillary shoot proliferation. World J Agric Sci 2:188–192

Sim GE, Goh CJ, Loh CS (1989) Micropropagation of *Citrus mitis* Blanco- multiple bud formation from shoot and root explants in the presence of 6- benzyaminopurine. Plant Sci 59:203–210

Simoes C, Albarello N, Callado CH, Carvalho de Castro T, Mansur E (2009) New approaches for shoot production and establishment of *in vitro* root cultures of *Cleome rosea* Vahl. Plant Cell Tiss Org Cult 98:79–86.

Singh A, Reddy MP, Patolia JS (2008) An improved protocol for micropropagation of elite genotypes of *Simmondsia chinensis* (Link) Schneider. Biol Plant 52:538–542

Singh KK, Gurung B (2009) *In vitro* propagation of *R. maddeni* Hook.F.an endangered *Rhododendron* species of Sikkim Himalaya. Not Bot Hort Agrobot Chij 37:79–83

Singh S, Rai M, Asthana P, Pandey S, Jaiswal V, Jaiswal U (2009) Plant gregeneration from alginate-encapsulated shoot tips of *Spilanthes acmella* L. Murr., a medicinally important and herbal pesticidal plant species. Acta Physiol Plant 31:649–894

Singh SK, Rai MK, Asthana P, Sahoo L (2010) Alginate-encapsulation of nodal segments for propagation, short-term conservation and germplasm exchange and distribution of *Eclipta alba* (L.) Acta Physiol Plant 32:607–610

Siril EA, Dhar U (1997) Micropropagation of mature Chinese tallow tree (*Sapium sebiferum* Roxb.). Plant Cell Rep 16:637–640

Sivakumar G, Kim SJ, Hahn EJ, Paek KY (2005) Optimizing environmental factors for large-scale multiplication of Chrysanthemum (*Chrysanthemum grandiflorum*) in balloon-type bioreactor culture. In Vitro Cell Dev Biol-Plant 41:822–825

Sivanesan I, Lee YM, Song JY, Jeong BR (2007) Adventitious shoot regeneration from leaf and petiole explants of *Campanula punctata* Lam and Rubriflora makino. Prop Ornament Plants 7:210–215

Skoog F, Miller CO (1957) Chemical regulation of growth and organ formation in plant tissues cultured *in vitro*. Symp. Soc Exp Biol 11:118–131

Skoog F, Tsui C (1951) Growth substances and the formation of buds in plant tissues. In: Skoog F (ed) Plant growth substances. University of Wisconsin Press, Madison, p 263–285

Smeets K, Cuypers A, Lambrechts A, Semane B, Hoet P, Van Laere A, Vangronsveld J (2005) Induction of oxidative stress and antioxidative mechanisms in *Phaseolus vulgaris* after Cd application. Plant Physiol Biochem 43:437–444

Standardi A, Piccioni E (1998) Recent perspectives on synthetic seed technology using nonembryogenic *in vitro* derived explants. Plant Sci 159:968–978

Stimart D, Harbage JH (1993) Growth of rooted "Gala" apple microcuttings *ex vitro* as influenced by initial adventitious root count. Hort Sci 28:664–666

Strnad M, Hanus J, Vanek T, Kaminek M, Ballantine JA, Fussel B, Hanke DE (1997) Meta-topolin, a highly active aromatic cytokinin from poplar leaves (*Populus* x *Canadensis* Moench., cv. Robusta). Phytochem 45:213–218

Stuart R, Street HE (1971) Studies on the growth in culture of plant cells. J Exp Bot 22:96–106

Sudha CG, Seeni S (1996) *In vitro* propagation of *Rauwolfia micrantha*, a rare medicinal plant. Plant Cell Tiss Org Cult 44:243–248

Sugiyama M (1999) Organogenesis *in vitro*. Curr Opin Plant Biol 2:61–64

Sujatha K, Hazra S (2007) Micropropagation of mature *Pongamia pinnata* Pierre. In Vitro Cell Dev Biol-Plant 43:608–613

Sukartiningsih, Nakamura K, Ide Y (1999) Clonal propagation of *Gmelina arborea* Roxb. by *in vitro* culture. J Fores Res 4:47–51

Sun Y, Zhao Y, Wang X, Qiao G, Chen G, Yang Y, Zhou J, Jin L, Zhuo R (2009) Adventitious bud regeneration from leaf explants of *Platanus occidentalis* L. and genetic stability assessment. Acta Physiol Plant 31:33–41

Sundararaj SG, Agrawal A, Tyagi RK (2010) Encapsulation for *in vitro* short-term storage and exchange of ginger (*Zingiber officinale* Rosc.) germplasm. Sci Hortic 125:761–766

Suttle GR (2000) Commercial laboratory production. In: Trigiano RN, Gray DJ (eds) Plant tissue culture concepts and laboratory exercises. CRC Press, Boca Raton, p 407–416

Swarup R, Parry G, Graham N, Allen T, Bennett M (2002) Auxin cross-talk: integration of signaling pathways to control plant development. Plant Mol Biol 49:411–426

Taha HS, Behkeet SA, Saker MM (2001) Factors affecting *in vitro* multiplication of date palm. Biol Plant 44:431–433

Tang W, Guo Z (2001) *In vitro* propagation of loblolly pine via direct somatic organogenesis from mature cotyledons and hypocotyls. Plant Growth Regul 33:25–31

Tereso S, Miguel C, Maroco J, Oliveira MM (2006) Susceptibility of embryogenic and organogenic tissues of maritime pine (*Pinus pinaster*) to antibiotics used in *Agrobacterium*-mediated genetic transformation. Plant Cell Tiss Org Cult 87:33–40

Thakur A, Kanwar JS (2008) Micropropagation of 'Wild Pear' *Pyrus pyrifolia* (Burm. F.) Nakai. I. explant establishment and shoot multiplication. Not Bot Hort Agrobot Cluj 35:103–108

Thomas T (2003) Advances in mulberry tissue culture. J Plant Biol 45:7–21

Thomas TD, Puthur JT (2004) Thidiazuron induced high frequency shoot organogenesis in callus from *Kigelia pinnata* L. Bot Bull Acad Sin 45:307–313

Thorpe T, Stasolla C, Yeung EC, De Klerk GJ, Roberts A, George EF (2008) The components of plant tissue culture media II: organic additions, osmotic and pH effects and support systems. In: George EF, Hall MA., De Klerk G-J. (edn) Plant propagation by tissue culture 3rd edition, vol.1. The background. Published by Springer, Dordrecht, p 143

Thorpe TA (ed) (1995) *In vitro* embryogenesis in plants. Kluwer, Dordrecht

Tiwari KP, Sharma MC, Tiwari SK (1997) Tissue culture protocols for teak (*Tectona grandis*), neem (*Azadirachta indica*) and khamer (*Gmelina arborea*). SFRI Technical Bulletin No. 29, State Forest Research Institute, Jabalpur

Tiwari SK, Tiwari KP, Siril EA (2002) An improved micropropagation protocol for teak. Plant Cell Tiss Org Cult 71:1–6

Torrey JG (1958) Differential mitotic response of diploid and polyploidy nuclei to auxin and kinetin treatment. Sci 128:1148

Tripathi M, Kumari N (2010) Micropropagation of a tropical fruit tree *Spondias mangifera* Willd. through direct organogenesis. Acta Physiol Plant 32:1011–1015

Tsay HS, Lee CY, Agrawal DC, Basker S (2006) Influence of ventilation closure, gelling agent and explant type on shoot bud proliferation and hyperhydricity in *Scrophularia yoshmurae*—a medicinal plant. Plant 42:445–449

Tsumara SE, Maschinski J, Keim P (1996) An analysis of genetic variation in *Astragalus cremnophylax* var. *cremnophylax*, a critically endangered plant using AFLP markers. Mol Ecol 5:735–745

Tsvetkov I, Hausman JE (2005) *In vitro* regeneration from alginate encapsulated microcuttings of *Quercus* sp. Sci Hort 103:503–507

Tsvetkov I, Huasman JF, Jouve L (2007) Thidiazuron-induced regeneration in root segments of White Popular (*P. alba* L.). Bulgarian J Agri Sci 13:623–626

Van Huylenbroeck J, Debergh PC (1996) Impact of sugar concentration *in vitro* on photosynthesis and carbon metabolism during *ex vitro* acclimatization of *Spathiphyllum* plantlets. Phys Plant 96:298–304

Van Huylenbroeck JM, Piqueras A, Debergh PC (2000) The evolution of photosynthesis capacity and the antioxidant enzymatic system during acclimatization of micropropagated *Calathea* plants. Plant Sci 155:59–66

Van Huylenbroeck JM, Van Laere IMB, Piqueras A, Debergh PC, Buneo P (1998b) Time course of Catalase and Superoxide dismutase during acclimatization and growth of micropropagated *Calathea* and *Spathiphyllum* plants. Plant Growth Reg 26:7–14

Van Staden J, Zazimalova E, George EF (2008) Plant growth regulators II: Cytokinins, their analogues and antagonists. In: George EF, Hall MA, De Klerk GJ (eds) Plant propagation by tissue culture, 3rd edn. Springer, Dordrecht, p 205–226

Varshney A, Anis M (2011) Improvement of shoot morphogenesis *in vitro* and assessment of changes of the activity of antioxidant enzymes during acclimation of micropropagated plants of Desert Teak. Acta Physiol. Plant. (In Press). DOI 10.1007/s11738-011-0883-9

Varshney A, Lakshmikumaran M, Srivastava PS, Dhawan V (2001) Establishment of genetic fidelity of in vitro-raised *Lilium* bulblets through RAPD markers. In Vitro Cell Dev Biol-Plant 37:227–231

Vasil IK (1959) Nucleic acids and the survival of excised anthers *in vitro*. Science 129:1487–1488

Vasil IK (ed) (1999) Advances in cellular and molecular biology of plants. Molecular improvement of cereal crops, vol 5. Kluwer, Dordrecht

Vejsadova H (2008) Growth regulator effect on *in vitro* regeneration of rhododendron cultivars. Hort. Sci. (Prague) 35:90–94

Vengadesan G, Selvaraj N, Amutha S, Ganapathi A (2002) *In vitro* propagation of *Acacia* species—a review. Plant Sci 163:663–671

Venkatachalam L, Sreedha RV, Bhagyalakshmi N (2007) Micropropagation in banana using high levels of cytokinins does not involve any genetic changes as revealed by RAPD and ISSR markers. Plant Growth Regul 51:193–205

Viana da Silva C, Silva de Oliveira L, Loriato VAP, Campos da Silva L, Salabert de Campos JM, Viccini LF, Jardim de Oliveira E, Otoni WC (2011) Organogenesis from root explants of commercial populations of *Passiflora edulis* Sims and a wild passionfruit species, *P. cincinnata* Masters. Plant Cell Tiss Org Cult 107:407–416

Vila S, Gonzalez A, Rey H, Mroginski L (2005) Plant regeneration, origin, and development of shoot buds from root segments of *Melia azedarach* L. (Meliaceae) seedlings. In Vitro Cell. Dev Biol- Plant. 41:746–751

Virchow R (1858) Die Cellullarpathologie im ihrer Begru¨ngung und physiologische und pathologische Gewebelehre. A Hirschwald, Berlin

Vochting H (1878) Uber Oganbildung im Pflanzenreich. Max Cohen, Bonn

Von Willert DJ, Matyssek R, Herppich W (1995) Experimentelle Pflanzenokologie. Gundlagem and Anwendungen, New York

Vos P, Hogers R, Blecker M, Reujans M, Van de Le T, Hornes M, Frijiters A, Pot A, Pelemen J, Kuiper M, Zabeau M (1995) AFLP: a new technique for DNA fingerprinting. Nucleic Acids Res 23:4407–4414

Wang HM, Zu YG, Dong FL, Zhao XJ (2005) Assessment of factors affecting *in vitro* shoot regeneration from axillary bud explants of *Camptotheca acuminata*. J Fores Res 16:52–54

Wang Q (1991) Shoot multiplication of pear in double-phase medium culture. Acta Hortic 289:349–350

Welsh J, McClelland M (1991) Genomic fingerprinting using arbitrarily primed PCR and a matrix of pair-wise combinations of primers. Nucleic Acids Res 19:5275–5279

Went FW (1928) Wuchstoff und Wachstum. Rec Trav Bot Neerl 25:1–116

Went FW, Thimann KV (1937) Phytohormones. The MacMillan Co., New York

Werbrouck SPO, Strnad M, Van Onckelen HA, Debergh PC (1996) Meta-topolin, an alternative to benzyladenine in tissue culture? Physiol. Plant 98:291–297

West TP, Ravindra MB, Preece JE (2006) Encapsulation, cold storage and growth of *Hibiscus moscheutos* nodal segments. Plant Cell Tiss Org Cult 87:223–231

White PR (1934a) Potentially unlimited growth of excised tomato root tips in a liquid medium. Plant Physiol 9:585–600

White PR (1934b) Multiplication of the viruses of tobacco and Aucuba mosaics in growing excised tomato root tips. Phytopath 24:1003–1011

White PR (1939) Potentially unlimited growth of excised plant callus in an artificial nutrient. Am J Bot 26:59–64

White PR (1942) Plant tissue cultures. Ann. Rev. Biochem. 11:615–628

White PR (1943) A handbook of plant tissue culture. Jacques Cattell Press, Tempe

Williams JGK, Kubelik AR, Livak KJ, Rafalski JA, Tingey SV (1990) DNA polymorphisms amplified by arbitrary primres are useful as genetic markers. Nucleic Acids Res 18:6531–6535

Williamson B, Cooke DEL, Duncan JM, Leifert C (1998) Fungal infections of micropropagated plants at weaning: a problem exemplified by downy mildews in *Rubus* and *Rosa*. Plant Cell Tiss Org Cult 52:89–96

Winton L (1968) Plantlets from asper tissue cultures. Sci 160:1234–1235

Wolter KE (1968) Root and shoot initiation in aspen callus cultures. Nature 219:509–510

Xu J, Yuzhen W, Zhang Y, Chai T (2008) Rapid *in vitro* multiplication and *ex vitro* rooting of *Malus zumi* (Matsumura) Rehd. Acta Physiol Plant 30:129–232

Yan H, Liang C, Yang L, Li Y (2010) *In vitro* and *ex vitro* rooting of *Siratia grosvenorii*, a traditional medicinal plant. Acta Physiol Plant 32:115–120

Yancheva SD, Golubowicz S, Fisher E, Lev-Yadun S, Flaishman MA (2003) Auxin type and timing of application determine the activation of the developmental program during in vitro organogenesis in apple. Plant Sci 165:299–309

Yannarelli GG, Gallego SM, Tomaro ML (2006) Effect of UV-B radiation on the activity and isoforms of enzymes with peroxidase activity in sunflower cotyledons. Environ Exp Bot 56:174–181

Yasodha R, Sumathi R, Gurumurthi K (2004) Micropropagation for quality propagule production in plantation forestry. Indian J Biotechnol 3:159–170

Yeo DY, Reed BM (1995) Micropropagation of three *Pyrus* rootstocks. HortSci 30:620–623

References

Yue D, Gosselin A, Desjardins Y (1993) Re-examination of the photosynthetic capacity of *in vitro*-cultured strawberry plantlets. J Am Soc Hort Sci 118:419–424

Zhihui S, Tzitzikas M, Raemakers K, Zhengqiang M, Visser R (2009) Effect of TDZ on plant regeneration from mature seeds in pea (*Pisum sativum*). In Vitro Cell Dev Biol- Plant 45:776–782

Zietkiewicz E, Rafalski A, Labuda D (1994) Genome fingerprinting by simple sequence repeat (SSR)-anchored polymerase chain reaction amplification. Genomics 20:176–183

Zimmerman RH, Fordham I (1985) Simplified method for rooting apple cultivars *in vitro*. J Am Soc Hortic Sci 110:34–38

Zobayed SMA, Saxena PK (2003) *In vitro*-grown roots: a superior explants for prolific shoot regeneration of St. John's wort (*Hypericum perforatum* L. cv. "New Stem") in a temporary immersion bioreactor. Plant Sci 165:463–470

Materials and Methods

Abstract

Plant tissue culture is the science of growing plant cells, tissues or organs excised from the mother/donor plant, on artificial media under an aseptic environment. It includes techniques and methods used to do research in many plant science disciplines and has several practical objectives. Before starting to culture plants via in vitro methods, it is necessary to have a comprehensible understanding of the ways in which plant material can be grown and manipulated in an in vitro environment. This chapter describes the techniques that have been developed for the micropropagation (clonal propagation) of the tree species *Balanites aegyptiaca* and shows these techniques can be applied for propagation of other tree species also. Both direct axillary shoot regeneration and adventitious regeneration are possible for clonal propagation in vitro.

3.1 Plant Materials and Explant Source

Mature seeds and nodal segments obtained from a 10-year-old candidate plus tree growing at the Arid Forest Research Institute, Jodhpur, India; juvenile cotyledonary nodes (15 days old); nodal segments (4 weeks old); and root explants (4 weeks old) collected from aseptic seedlings were used in this study.

3.2 Surface Disinfection of Seeds and Explants

The seeds from mature fruits and mature nodal explants were thoroughly washed in running tap water for 30 min to remove adherent particles, immersed in 1 % (w/v) solution of Bavistin, a fungicide for 30 min, then treated with 5 % (v/v) Teepol solution for 20 min and rinsed thrice with sterile distilled water followed by surface sterilization in 0.1 % (w/v) mercuric chloride solution under sterile conditions for 10 min for seeds and 5 min for nodal explants and then rinsed thrice with sterile distilled water. The disinfected seeds and mature nodal explants (2 cm) were inoculated onto respective sterile culture media (germination or onto the shoot induction medium).

3.3 Establishment of Aseptic Seedlings and Preparation of Explants

The surface-sterilized seeds were germinated on Murashige and Skoog (1962) basal medium alone or supplemented with gibberellic acid (GA_3) at

Table 3.1 Composition of MS (Murashige and Skoog), B_5 and L_2 media

Components	MS	B_5	L_2
Macronutrients		Amount (mg/l)	
$MgSO_4$	370	250	435
KH_2PO_4	170	–	325
KNO_3	1,900	2,500	2,100
NH_4NO_3	1,650	–	1,000
$CaCl_2 \cdot 2H_2O$	440	150	600
$NaH_2PO_4 \cdot H_2O$	–	150	85
$(NH_4)_2SO_4$	–	134	–
Micronutrients			
H_3BO_3	6.2	3.0	5.0
$MnSO_4 \cdot 4H_2O$	22.3	–	19.8
$MnSO_4 \cdot H_2O$	–	10.0	–
$ZnSO_4 \cdot 7H_2O$	8.6	2.0	5.0
$Na_2MoO_4 \cdot 2H_2O$	0.25	0.25	0.4
$CuSO_4 \cdot 5H_2O$	0.025	0.025	0.1
$CoCl_2 \cdot 6H_2O$	0.025	0.025	0.1
KI	0.83	0.75	1.0
$FeSO_4 \cdot 7H_2O$	27.8	27.8	25.0
$Na_2EDTA \cdot 2H_2O$	37.3	37.3	33.5
Organic supplements			
Thiamine HCl	0.5	10.0	2.0
Pyridoxine HCl	0.5	1.0	0.5
Nicotinic acid	0.5	1.0	–
Myo-inositol	100	100	250
Glycine	20	–	–

different concentrations (0.5, 1.0, 2.0, 2.5 and 5 µM) to raise aseptic seedlings. Nodal segments (1–1.5 cm), cotyledonary nodes (1–1.5 cm) and root segments (1–1.5 cm) from aseptic seedlings were used as explants.

3.4 Culture Media

Murashige and Skoog (MS; Murashige and Skoog 1962) medium was used as primary basal medium for in vitro studies. Other media like B_5 (Gamborg et al. 1968), L_2 (Phillips and Collins 1979) and woody plant medium (WPM; Lloyd and McCown 1980) were also tested for maximum shoot induction and proliferation.

3.4.1 Composition of Basal Media

The different constituents of MS, B_5 and L_2 media along with their concentrations used are listed in Table 3.1. WPM medium was used as a ready-made preparation procured from the Duchefa Biochemei Company.

3.4.2 Preparation of Stock Solutions

The constituents of MS, B_5 and L_2 depicted in Table 3.2 were prepared in the form of four different stock solutions: stock I—major salts (20×), stock II—minor salts (200×), stock III—$FeSO_4 \cdot 7H_2O$ and $Na_2EDTA \cdot 2H_2O$ (100×), and stock IV—organic nutrients except sucrose (100×).

Table 3.2 Stock solutions for MS (Murashige and Skoog), B_5 and L_2 media

Components	MS	B_5	L_2
Stock solutions I (20×)		Amount (mg/l)	
$MgSO_4$	7,400	5,000	8,700
KH_2PO_4	3,400	–	6,500
KNO_3	38,000	50,000	42,000
NH_4NO_3	33,000	–	20,000
$CaCl_2·2H_2O$	8,800	3,000	12,000
$NaH_2PO_4·H_2O$	–	3,000	1,700
$(NH_4)_2SO_4$	–	2,680	–
Stock solution II (200×)			
H_3BO_3	1,240	600	1,000
$MnSO_4·4H_2O$	4,460	–	3,960
$MnSO_4·H_2O$	–	2,000	–
$ZnSO_4·7H_2O$	1,720	400	1,000
$Na_2MoO_4·2H_2O$	50	50	80
$CuSO_4·5H_2O$	5.0	5.0	20
$CoCl_2·6H_2O$	5.0	5.0	20
KI	166	150	200
Stock solution III (100×)			
$FeSO_4·7H_2O$	2,780	2,780	2,500
$Na_2EDTA·2H_2O$	3,730	3,730	3,350
Stock solution IV (100×)			
Thiamine HCl	50	1,000	200
Pyridoxine HCl	50	100	50
Nicotinic acid	50	100	–
Myo-inositol	10,000	10,000	25,000
Glycine	200	–	–

All stock solutions were made separately by dissolving the required amount of solute in double distilled water (DDW). The reasons for preparing different stock solutions is that certain kinds of chemicals, when mixed together, will precipitate and do not remain in solutions. To prepare 1 litre of medium, 50 ml of stock solution I, 5 ml of stock solution II and 10 ml of each of stock solution III and IV were taken. Separate stock solutions were prepared for each plant growth regulator by dissolving it in a minimal quantity of the appropriate solvent (1 N NaOH or absolute alcohol) and making up to the desired volume with DDW. All the stock solutions were stored in a refrigerator at 4 °C and were checked properly before use.

3.5 Plant Growth Regulators

Depending upon the experimental setup, the basal medium was supplemented with various plant growth regulators such as cytokinins viz., 6-benzyladenine (BA), 6-furfurylaminopurine/kinetin (Kn), 2-isopentenyladenine (2-iP) or thidiazuron (TDZ) either singly at various concentrations (1.0, 2.5, 5.0, 10.0, 12.5 and 15.0 μM) or in combination with auxins viz., indole-3-butyric acid (IBA), indole-3-acetic acid (IAA) and α-naphthalene acetic acid (NAA) at different concentrations (0.5, 1.0, 2.0 and 2.5 μM).

3.6 Adjustment of pH, Gelling of Medium, Carbon Source and Sterilization

The pH of the MS, WPM and L_2 media were adjusted to 5.8 and B_5 to 5.5 using 1N NaOH or HCl using a pH meter (Elico Pvt. Ltd., India) prior to autoclaving. The medium was solidified with 0.8% (w/v) agar (Qualigen's Fine Chemicals, Mumbai) by dissolving it in a microwave until a clear gel is formed.

The experiments were also designed to assess the effect of different pH values (5.0, 5.4, 5.3, 6.0 and 6.4) at 3% sucrose (sole carbon source) and different sucrose concentrations [1, 2, 3, 4, 5 and 6% (w/v)] at pH 5.8 of the medium with optimal dose of the plant growth regulator regimen for maximum shoot regeneration.

The media was dispensed in 25×150 mm test tubes each containing 20 ml of medium while 50 ml in 100-ml capacity Erlenmeyer flasks (Borosil, India) and cotton plugs (single-layered cheesecloth stuffed with non-absorbent cotton) were used as closures. The medium was sterilized in an autoclave at 1.06 kg cm^{-2} (121 °C) for 15 min and the medium in culture tubes were allowed to set as slants.

3.7 Sterilization of Glassware and Instruments

All the glassware, instruments (wrapped in aluminium foil) and DDW were sterilized by autoclaving at 1.06 kg cm^{-2} (121 °C) for 20 min. The forceps, scalpel, etc. made of stainless steel were sterilized by dipping them in rectified spirit followed by flaming and cooling before inoculation.

3.8 Sterilization of Laminar AirFlow Hood

The laminar airflow cabinet (NSW, Delhi) was sterilized by switching on ultraviolet (UV) light for 30 min followed by wiping the working surface area with 70% alcohol before any operation inside the cabinet.

3.9 Inoculation and Culture Conditions

Inoculation was performed under aseptic conditions of the laminar airflow cabinet by using sterilized culture media, instruments and distilled water. The instruments were re-sterilized (from time to time) during inoculation by dipping them in absolute alcohol followed by their flaming and cooling. The surface-sterilized plant materials were transferred to petri dishes and inoculated using sterilized forceps in culture vials followed by plugging with cotton plugs in quick succession.

All the cultures were maintained at 24 ± 2 °C under a 16/8-h photoperiod with a photosynthetic photon flux density (PPFD) at 50 µmol m^{-2} s^{-1} provided by cool white fluorescent light (40 W, Philips, India) with 55–60% relative humidity.

3.10 Rooting

Rooting was attempted in the microshoots (3–4 cm) using in vitro and ex vitro methods. For in vitro root induction, the isolated shoots were transferred on full- and half-strength MS media with or without auxins (IAA, IBA, NAA) at various concentrations (0.1, 0.5, 1.0, 2.0 and 5.0 µM).

For ex vitro rooting, the basal end of the healthy shoots was dipped in IBA solution at concentrations of 100, 150, 200, 250 and 300 µM for 30 min and then planted in small thermocol cups containing sterile soilrite (Keltech Energies Pvt. Ltd.) and acclimatized according to the procedure described below.

3.11 Acclimatization and Hardening of Plantlets

Plantlets with a well-developed shoot and root system were removed from the culture medium and washed gently under running tap water to remove any adherent gel from the roots and transferred to thermocol cups containing sterile vermiculite, soilrite (Keltech Energies Pvt. Ltd.,

India) or garden soil. These were kept under diffuse light conditions (16:8h photoperiod) covered with transparent polythene bags to ensure high humidity, irrigated every 3 days with half-strength MS salt solution (without vitamins) for 2 weeks. The polythene membranes were removed after 4 weeks in order to acclimatize plantlets, and after 8 weeks they were transferred to pots containing garden soil or a mixture of garden soil and vermicompost (1:1) and maintained in a greenhouse under normal day-length conditions.

3.12 Synthetic Seed

3.12.1 Explant Source

Nodal segments (1.0 cm), excised from in vitro shoot culture, were used as explants.

3.12.2 Encapsulation Matrix

Sodium alginate (Qualigens, India) at different concentrations (2, 3, 4 and 5%) was added to liquid MS medium. For complexation, 25, 50, 75, 100 and 200 mM $CaCl_2 \cdot H_2O$ solution was prepared using liquid MS medium. The pH of the gel matrix and the complexing agent was adjusted to 5.8 prior to autoclaving at 121 °C for 20 min.

3.12.2.1 Encapsulation
Encapsulation was accomplished by mixing the nodal segments in sodium alginate solution and dropping them in $CaCl_2 \cdot 2H_2O$ solution using a pipette. The droplets containing the explants were held at least for 20 min to achieve polymerization to form beads. The calcium alginate beads containing the nodal segments were retrieved from the solution with a tea strainer, rinsed twice with autoclaved distilled water to remove the traces of $CaCl_2 \cdot 2H_2O$, transferred to sterile filter paper in petri dishes for 5 min under the laminar airflow cabinet to eliminate the excess of water and thereafter planted onto petri dishes or tubes containing the nutrient medium.

3.12.2.2 Planting Media
The encapsulated nodal segments (alginate beads) were transferred to wide-mouth culture flasks (Borosil, India) containing MS basal medium and MS medium supplemented with plant growth regulators either singly or in combination as specified in the results. The alginate beads were then incubated under the same culture conditions as specified in the section 'Inoculation and culture conditions' of this chapter.

3.12.2.3 Low-temperature Storage
A set of encapsulated nodal segments were transferred to water and agar medium and stored in the refrigerator at 4 °C. Seven different exposure times (0, 1, 2, 3, 4, 5 and 6 weeks) at low temperature were evaluated for regeneration. After each storage period, encapsulated nodal segments were cultured on half-strength MS medium supplemented with plant growth regulators for conversion into plantlets. The percent conversion of encapsulated nodal segments onto regeneration medium was recorded after 4 weeks of culture. The plantlets developed from encapsulated nodal segments were hardened off and acclimatized as specified in the results.

3.13 Physiological and Biochemical Studies of in vitro Regenerated Plants During Acclimatization

3.13.1 Chlorophyll and Carotenoid Estimation

The chlorophyll (Chl a and b) and carotenoid content of leaf tissue was estimated using the method of Mackinney (1941) and Maclachan and Zalick (1963), respectively. About 100 mg of fresh leaf tissues were ground in 5 ml acetone (80%) with the help of a mortar and pestle. The suspension was filtered with Whatman filter paper number-1, if necessary the supernatant was again washed and filtered, the total filtrate was taken in graduated test tubes and the final volume was made up to 10 ml with 80% acetone. The optical densities (ODs) of the chlorophyll solution were read at wavelengths 645 and 663 nm, and for carotenoids, the OD was read at 480 and 510 nm wavelengths with the help of a spectrophotometer (UV-Pharma Spec 1600, Shimadzu, Japan).

The chlorophyll (Chl a and b) and carotenoid content was expressed in mg/g fresh tissue and calculated according to the following formulae:

Chlorophyll a
$$= \frac{12.7(\text{OD 663 nm}) - 2.69(\text{OD 645 nm}) \times V}{d \times 1000 \times W}$$

Chlorophyll b
$$= \frac{22.9\ (\text{OD 645 nm}) - 4.68(\text{OD 663 nm}) \times V}{d \times 1{,}000 \times W},$$

Carotenoids
$$= \frac{7.6(\text{OD 480 nm}) - 1.49(\text{OD 510 nm}) \times V}{d \times 1{,}000 \times W},$$

where
- V = final volume of chlorophyll extract in 80% acetone
- d = length of light path
- W = fresh weight of leaf tissue
- OD = optical density at the given wavelength.

3.14 Assessment of Antioxidant Enzyme Activities

3.14.1 Superoxide Dismutase

Superoxide dismutase (SOD; superoxide to superoxide oxidoreductase; EC 1.15.1.1) activity was measured as given by Dhindsa et al. (1981) with slight modifications.

3.14.1.1 Preparation of Reagents

- **1M Sodium bicarbonate solution**

15.9 g of sodium bicarbonate was dissolved in distilled water and the volume was made up to 100 ml.

- **200 mM Methionine solution**

2.98 g of methionine was dissolved in DDW and the volume was made up to 100 ml.

- **2.25 mM Nitroblue tetrazolium solution**

0.184 g of nitroblue tetrazolium (NBT) was dissolved in 100 ml of DDW.

- **3 mM EDTA**

It was prepared by dissolving 1.116 mg of ethylenediaminetetraacetic acid (EDTA) in 100 ml DDW.

- **60 µM Riboflavin**

It was prepared by dissolving 2.3 mg riboflavin in 100 ml of DDW.

- **Extraction buffer**
- **0.5 M Potassium phosphate buffer (pH 7.3)**

It was prepared from 0.5 M phosphate buffer (pH 7.3). The solution of monobasic potassium phosphate (KH_2PO_4) and dibasic potassium phosphate (K_2HPO_4) was first prepared in the following manner:

Solution A 3.40 g of KH_2PO_4 was dissolved in DDW and the volume was made up to 50 ml.

Solution B 8.70 g of K_2HPO_4 was dissolved in DDW and the volume was made up to 100 ml.

To prepare the extraction buffer, solutions A and B were mixed in an appropriate ratio and the pH was adjusted to 7.3 with the help of a pH meter. To 100 ml of this buffer, 1 g of polyvinylpyrrolidone (PVP), 1 ml of Triton X-100 and 0.11 g of EDTA were added.

- **Reaction buffer**
- **0.1M Potassium phosphate buffer (pH 7.8)**

A phosphate buffer of 0.1 M (pH 7.8) was used as an extraction buffer. The solutions of potassium dihydrogen phosphate (KH_2PO_4) and dipotassium hydrogen phosphate (K_2HPO_4) were prepared in the following manner.

Solution A 1.30 g of KH_2PO_4 was dissolved in DDW and the volume was made up to 50 ml.

Solution B 1.70 g of K_2HPO_4 was dissolved in DDW and the volume was made up to 100 ml.

Solutions A and B were mixed in an appropriate ratio to adjust the pH at 7.8 with the help of a pH meter. To 100 ml of this buffer, 1.0 g of PVP was added.

3.14.1.2 Enzyme Extraction and Assay

Two hundred milligrams of fresh leaf samples of 0-, 7-, 14-, 21- and 28-day-old micropropagated plants were homogenized in 2.0 ml of extraction buffer containing 1% PVP, 1% Triton X-100 and 0.11 g of EDTA using a pre-chilled mortar and pestle. The process of homogenization was carried out in an icebox at 4 °C. The homogenate was transferred to centrifuge tubes and centrifuged at 12,000 rpm for 15 min at 4 °C.

The SOD activity in the supernatant was assayed by its ability to inhibit the photochemical reduction of NBT. The assay mixture, consisting of 1.5 ml reaction buffer, 0.2 ml of methionine, 0.1 ml enzyme extract with equal amount of 1M Na_2HCO_3, 2.25 mM NBT solution, 3 mM EDTA, riboflavin and 1.0 ml of DDW, was taken in test tubes which were incubated under the light of a 15-W fluorescent lamp for 10 min at 25/28 °C. Blank A containing all the above substances of the reaction mixture, along with the enzyme extract, was placed in light for 10 min and then under dark conditions. Blank B containing all the above substances of reaction mixture except enzyme extract was placed in light along with the sample. The reaction was terminated by switching off the light, and the tubes were covered with a black cloth. The non-irradiated reaction mixture containing enzyme extract did not develop light blue colour. Absorbance of samples along with blank B was read at 560 nm against the blank A. The difference of percent reduction in the colour between blank B and the sample was then calculated. The enzyme activity was expressed in enzyme units (EU) mg^{-1} protein.

3.14.2 Catalase

Catalase (CAT; H_2O_2 to H_2O_2 oxidoreductase; EC 1.11.1.6) activity in the leaves was determined by the method of Aebi (1984) with slight modifications.

3.14.2.1 Preparation of Reagents
- **3mM H_2O_2**

0.1 ml of H_2O_2 was mixed with 9.9 ml of DDW.
- **3mM EDTA**

1.116 mg was dissolved in DDW and the volume was made up to 100 ml.
- **Extraction Buffer**
- **0.5 M Potassium Phosphate Buffer (pH 7.3)**

Solution A 3.40 g of KH_2PO_4 was dissolved in DDW and the volume was made up to 50 ml.

Solution B 8.70 g of K_2HPO_4 was dissolved in DDW and the volume was made up to 100 ml.

Solutions A and B were in an appropriate ratio and pH was adjusted to 7.3 with the help of a pH meter. To 100 ml of this buffer, 1.0 g PVP, 1.0 ml Triton X-100 and 0.11 g of EDTA were added.
- **Reaction Buffer**
- **0.5 M/0.25 M Phosphate buffer (pH 7.2/7.0)**

Solution A 3.40/1.70 g of KH_2PO_4 was dissolved in DDW and the volume was made up to 50 ml.

Solution B 8.70/4.35 g of K_2HPO_4 was dissolved in DDW and the volume was made up to 100 ml.

Solutions A and B were in an appropriate ratio and the pH was adjusted to 7.2/7.0 with the help of a pH meter.

3.14.2.2 Enzyme Extraction and Assay

Five hundred milligrams of fresh leaf samples of 0-, 7-, 14-, 21- and 28-day-old micropropagated plants were homogenized in 5 ml of extraction buffer containing 1% PVP, 1% Triton X-100 and 0.11 g EDTA using a pre-chilled mortar and pestle. The process of homogenization was carried out in an icebox at 4 °C. The supernatant was used immediately for enzyme assay.

Catalase activity was determined by monitoring the disappearance of H_2O_2 by measuring a decrease in absorbance at 240 nm. The reaction was carried out in a final volume of 2 ml of the reaction mixture containing reaction buffer with 0.1 ml 3 mM EDTA, 0.1 ml of the enzyme extract and 0.1 ml of 3 mM H_2O_2. The reaction was allowed to run for 5 min. The activity was calculated by using the extinction coefficient (ε), 0.036 mM^{-1} cm^{-1}, and expressed in EU mg^{-1} protein. One unit of the enzyme determines the amount necessary to decompose 1 µmol of H_2O_2 per minute at 25 °C.

3.14.3 Ascorbate Peroxidase

Ascorbate peroxidase (APX; L-ascorbate to H_2O_2 oxidoreductase; EC 1.11.1.11) activity was determined by the method used by Nakano and Asada (1981).

3.14.3.1 Preparation of Reagents
- **0.5 mM Ascorbate**

44 mg of L-ascorbate was dissolved in DDW and the volume was made up to 100 ml.
- **0.3 % (v/v) H_2O_2**

1 ml of 30% H_2O_2 was mixed with 99 ml of DDW.
- **Extraction Buffer**
- **100 mM Potassium phosphate buffer (pH 7.8)**

Solution A 1.36 g KH_2PO_4 was dissolved in DDW and the volume was made up to 100 ml.

Solution B 1.74 g of K_2HPO_4 was dissolved in DDW and the volume was made up to 100 ml.

To prepare the extraction buffer, the two solutions A and B were mixed together in an appropriate amount and the pH was adjusted to 7.0 with the help of a pH meter. To 100 ml of this buffer, 1.0 g of PVP, 1.0 ml Triton X-100 and 0.11 g of EDTA were added.
- **Reaction Buffer**
- **50 mM Sodium phosphate buffer (pH 7.2)**

Solution A 1.142 g of NaH_2PO_4 was dissolved in DDW and the volume was made up to 100 ml.

Solution B 0.707 g of Na_2HPO_4 was dissolved in DDW and the volume was made up to 100 ml.

The buffer (pH 7.2) was prepared by mixing these two, A and B, solutions in an appropriate ratio and the pH was adjusted to 7.2 with the help of a pH meter.

3.14.3.2 Enzyme Extraction and Assay

Hundred milligrams of fresh leaf samples of 0-, 7-, 14-, 21- and 28-day-old micropropagated plants were ground in 4 ml of extraction buffer containing 1% PVP, 1% Triton X-100 and 0.11 g of EDTA. APX activity was determined by the decrease in the absorbance of ascorbate at 290 nm, due to its enzymatic breakdown; 1.0 ml of the reaction buffer, 0.1 ml of 0.3% H_2O_2, 0.1 ml of 0.5 mM ascorbate, 0.1 ml of 0.5 mM EDTA and 100 µl of the enzyme extract were used. The reaction was allowed to run for 5 min at 25 °C. APX activity was calculated by using its extinction coefficient (ε), 2.8 $mM^{-1} cm^{-1}$, and expressed in EU mg^{-1} protein. One unit of the enzyme determines the amount necessary to decompose 1 µmol of substrate consumed per minute at 25 °C.

3.14.4 Glutathione Reductase

Glutathione reductase (GR; NADPH to glutathione-disulphide oxidoreductase; EC 1.6.4.2) activity was determined by the method of Foyer and Halliwell (1976) as modified by Rao (1992).

3.14.4.1 Preparation of Reagents
- **Extraction buffer**
- **100 mM Potassium phosphate buffer (pH 7.0)**

Solution A 1.36 g KH_2PO_4 was dissolved in DDW and the volume was made up to 100 ml.

Solution B 1.74 g of K_2HPO_4 was dissolved in DDW and the volume was made up to 100 ml.

To prepare the extraction buffer, the two solutions A and B were mixed together in an appropriate amount and the pH was adjusted to 7.0 with the help of a pH meter. To 100 ml of this buffer, 1.0 g of PVP, 1.0 ml Triton X-100 and 0.11 g of EDTA were added.
- **Reaction Buffer (0.25 M/0.1 M Tris buffer)**
- **0.2 mM NADPH**

2 mg of NADPH was dissolved in 10 ml of DDW.

0.5 mM oxidized glutathione

4 mg of oxidized glutathione (GSSG) was dissolved in 13 ml of DDW.

3.14.4.2 Enzyme Extraction and Assay

Five hundred milligrams of fresh leaf samples of 0-, 7-, 14-, 21- and 28-day-old micropropagated plants were homogenized in 2 ml of extraction buffer containing 1% PVP, 1% Triton X-100 and 0.11 g of EDTA in a pre-chilled mortar and pestle. The process of homogenization was carried out in an icebox at 4 °C. The homogenate was centrifuged at 12,000 rpm for 15 min at 4 °C. The supernatant was collected and used for the enzyme assay. The GR activity was determined by monitoring the glutathione-dependent oxidation of NADPH at its absorption maxima of the wavelength 340 nm. The reaction mixture was prepared by adding 1.0 ml reaction buffer (0.2 mM NADPH, 0.5 mM GSSG) and 0.1 ml enzyme extract. The reaction was allowed to run for 5 min at 25 °C. Corrections were made for any GSSG oxidation in the absence of NADPH. The activity was calculated by using its extinction coefficient (ε), 6.2 mM^{-1} cm^{-1}, and expressed in EU mg^{-1} protein. One unit of the enzyme determines the amount necessary to decompose 1 μmol of NADPH per minute at 25 °C.

3.14.5 Soluble Protein

The total soluble protein content of the leaves of regenerants was estimated following the method of Bradford (1976) using bovine serum albumin (BSA, Sigma, USA) as the standard.

3.14.5.1 Preparation of Reagents

- **10% (w/v) TCA**

10 g of trichloroacetic acid (TCA) was dissolved in DDW to make a final volume of 100 ml.

- **0.1N NaOH solution**

0.4 g of NaOH pellets was dissolved in DDW to make a final volume of 100 ml.

- **Bradford's reagent**

Fifty millilitres of 90% ethanol was mixed to 100 ml of orthophosphoric acid (85%). Its volume was made up to 1 l and 100 mg of Coomassie Brilliant Blue (G) dye was added to it which was stirred well on a magnetic stirrer in a dark, covered volumetric flask. The solution was then filtered through Whatman filter paper number-1 and stored under dark conditions. The resultant reagent was called Bradford's reagent. The final concentrations of the components in the reagent were 0.01% Coomassie Brilliant Blue G-250 (w/v), 4.75% ethanol (w/v) and 8.5% O-phosphoric acid (w/v).

- **Extraction buffer**

A 0.1M phosphate buffer (pH 7.2) was used as extraction buffer. The solution of potassium dihydrogen phosphate (KH$_2$PO$_4$) and dipotassium hydrogen phosphate (KH$_2$PO$_4$) was prepared in the following manner:

Solution A 1.30 g of KH$_2$PO$_4$ was dissolved in DDW and the volume was made to 100 ml.

Solution B 1.70 g K$_2$HPO$_4$ was dissolved in DDW and the volume was made to 100 ml.

Solutions A and B were mixed in an appropriate ratio to adjust the pH at 7.2 with the help of a pH meter. To 100 ml of this buffer, 1.0 g PVP was added.

3.14.5.2 Extraction and Estimation of Total Soluble Protein

Five hundred milligrams of fresh leaf material were homogenized in 5 ml of 0.1 M phosphate buffer (extraction buffer) at 4 °C with the help of a pre-chilled mortar and pestle and kept in an icebox during the process of homogenization. The homogenate was transferred to a 30-ml centrifuge tube and centrifuged at 5,000 rpm for 10 min at 4 °C. An equal amount of chilled 10% TCA was added to 1 ml of the supernatant, which was again centrifuged at 3,300 rpm for 10 min. The supernatant was discarded and the pellets were washed with acetone. It was dissolved in 1 ml of 0.1 N NaOH.

To 0.1 ml of the aliquot, 0.5 ml of Bradford's reagent was added and mixed using a vortex mixer. The tubes were kept for 10 min for optimal colour development. The absorbance was then recorded at 595 nm on a UV-visible spectrophotometer. The soluble protein concentrations were quantified with the help of a standard curve prepared from the standard of bovine serum albumin (BSA) from Sigma, USA. The protein content was expressed in mg g^{-1} fresh weight.

3.15 Anatomical Studies

3.15.1 Fixation and Storage

The differentiating explants were fixed in FAA solution consisting of formalin to glacial acetic acid to alcohol (70%) in the ratio of 4:6:90 (v/v). The fixed samples were stored in 70% alcohol.

3.15.2 Embedding, Sectioning and Staining

The standard method of paraffin embedding (Johansen 1940) was followed for histological studies. The ethanol–xylol series was used for dehydration and infiltration. For complete infiltration, the plant material to be sectioned was kept in a vacuum oven at 60 °C for 15 min. Sections (longitudinal and transverse) of 10 µM thickness were cut using a Spencer 820 microtome (American Optical Corp., Buffalo, NY) and the resulting paraffin ribbons were passed through a series of deparafinizing solutions and stained in safranin and fast green solutions. Permanent slides were made by using Canada balsam. The sections were examined under a light microscope (Olympus CH20i, Japan).

3.16 Chemicals and Glassware Used

Most of the chemicals like BSA, EDTA, GR, glutathione (GSH), GSSG, PVP, Triton X-100, NBT, H_2O_2, methionine, TCA, plant growth regulators (BA, Kn, 2-iP, TDZ, IAA, IBA, NAA), NADH, riboflavin, etc. were obtained from Sigma-Aldrich Pvt. Ltd., New Delhi, India, and/or from Sigma Aldrich (St. Louis, MO, USA). Other major and minor salts and buffer components were procured from Qualigens, Merck and/or SRL. All chemicals were obtained in the highest purity available commercially.

Glassware, such as test tubes (25 mm × 150 mm), petri dishes (17 mm × 100 mm), wide-mouth Erlenmeyer flasks (100 and 250 ml), were purchased from Borosil, India.

3.17 Statistical Analysis

All experiments had ten replicates per treatment and each experiment was repeated thrice. Completely randomized block design (CRBD) was used to test the effects of different concentrations of cytokinins and auxins. Data were analysed statistically using one-way analysis of variance (ANOVA). The significance of differences among means were carried out using Duncan's multiple range test (DMRT) at $P = 0.05$ SPSS Ver. 16 (SPSS Inc., Chicago, IL, USA). The results are expressed as mean ± standard error (SE) of three experiments.

3.18 Genomic DNA Isolation and Purification

Genomic DNA was isolated from fresh leaf tissues of the selected micropropagated plants derived from mature nodal explants of donor plant and the control plant (represented as the outlier) using the cetyltrimethylammonium bromide (CTAB) method (Doyle and Doyle 1990) with slight modifications.

3.18.1 Preparation of Stock Solutions Required for DNA Extraction

- **100 ml 1M Tris–HCl (pH = 8.0)**

For preparing 100 ml of 1M Tris–HCl solution, 15.76 g Tris–HCl (Sigma, USA) was dissolved in 80 ml of DDW. The pH of the solution was adjusted to 8.0 by dropwise addition of concentrated HCl (Qualigens, India). After pH adjustment, the final volume of the solution was made to 100 ml with DDW.

- **100 ml 0.5 M EDTA (pH = 8.0)**

For preparing 100 ml of 0.5M EDTA solution, 14.6 g EDTA (Sigma, USA) was dissolved first in 50 ml of DDW. The pH of the solution was

3.18 Genomic DNA Isolation and Purification

adjusted to 8.0 by addition of NaOH pellets (Qualigens, India). After pH adjustment, the final volume of the solution was made to 100 ml with DDW.

- **5M NaCl solution**

For preparing 100 ml of 5M NaCl solution, 29.22 g NaCl (Sigma, USA) was dissolved in 100 ml of DDW.

- **2.5% CTAB solution**

For preparing 100 ml of 2.5% CTAB solution, 2.5 g of CTAB (Sigma, USA) was dissolved in 100 ml of DDW.

- **CTAB DNA Extraction Buffer**

The CTAB (Sigma, USA) extraction buffer was prepared by mixing the stock solution reagents in the desired amount in a sterilized conical flask as mentioned below:

- **Amount taken from the stock solution (10 ml)**

1.0 ml 1M Tris–HCl (pH = 8.0), 3.0 ml 5M NaCl, 500 µl 0.5M EDTA (pH = 8.0), 1.25 ml 2.5% CTAB, 4.23 ml mQ water

- **Tris–EDTA buffer (10 ml)**

This is composed of the following chemicals:
1M Tris–HCl (500 µl)
0.5M EDTA (200 µl)
mQ Water (9.3 ml)

3.18.2 Extraction and Purification Protocol

For the extraction of genomic DNA, 1 g of fresh leaf tissues was frozen using liquid nitrogen avoiding the thawing of samples. The frozen samples were ground thoroughly into a fine powder in a chilled mortar and pestle. After the complete homogenization of plant materials, 3 ml of the CTAB extraction buffer was added (containing 1.0% (w/v) PVP and 0.2% (v/v) β-mercaptoethanol), and the samples were homogenized again into a fine paste. The resultant mixture was then transferred to non-reactive polypropylene centrifuge tubes (15 ml) and incubated in a fixed temperature water bath at 65 °C for 1 h with occasional inversions after every 20 min. After incubation, equal volumes of chloroform to isoamylalcohol (24:1) was added to avoid any protein contamination, debris and interphase material. The solutions were mixed thoroughly and centrifuged at 15,000 rpm at 4 °C for 10 min.

The supernatant was collected and transferred to Eppendorf tubes containing 170 µl 5M sodium chloride solution and a double volume of chilled isopropanol (Qualigens, India). The resultant solution was mixed well by inversion three to five times and kept for 30 min at −20 °C. White fluffy DNA strands appeared in the Eppendorf tubes which were recovered by centrifugation at 10,000 rpm for 10 min at 4 °C. The supernatant was discarded and the white pellet at the bottom of the Eppendorf was air dried properly. The pellet was dissolved in 300 µl of Tris–EDTA (TE) buffer and incubated at 4 °C for 30 min. To the dissolved pellet, 3 µl of RNAse was added and incubated for 30 min at 37 °C to degrade the RNA from the genomic DNA. Next, 25 µl sodium acetate and 600 µl ethanol was added to the dissolved pellet and again incubated for 30 min at −20 °C. The dissolved pellet was centrifuged at 10,000 rpm for 10 min at 4 °C. The supernatant was discarded and the pellet was air dried and finally dissolved in 100 µl of mQ water.

3.18.3 Quantitative and Qualitative Assessment of Genomic DNA

3.18.3.1 Quantification

Isolated genomic DNA was quantified using a nanophotometer (Implen). The optical density (absorbance A) was measured at 260 nm (A_{260}) and 280 nm (A_{280}). The quantity of DNA present in the solution was calculated from absorption at 260 nm (A_{260}) and the purity of DNA was calculated from the A_{260}/A_{280} ratio. For an ideal preparation, the A_{260}/A_{280} ratio should be ≥ 1.8 (Sambrook et al. 2001).

3.18.3.2 Quality Analysis

The quality of the extracted DNA was tested by running the DNA in 1% agarose gel.

Solutions for Agarose Gel Electrophoresis
- **Running Buffer (1 × TBE)**

To prepare 500 ml of 5× TBE stock solution, 27 g Tris base (Sigma, USA),10 ml of 0.5 M EDTA (Sigma, USA) and 13.7 g of boric acid were dissolved in 250 ml of DDW. The pH of the solution was adjusted to 8.0 with the dropwise addition of concentrated HCl, and the final volume was made to 500 ml with DDW. During gel electrophoresis, the 5× TBE buffer was diluted with DDW to obtain 1× TBE buffer (working concentration) before running the gel.

- **Gel-Loading Dye (6×)**

To prepare 25 ml of the gel-loading dye (6×), 30 mg of bromophenol blue dye (Sigma, USA) was mixed with 9.36 ml of 80% glycerol (Qualigens, India), 30 mg xylene cyanol (Merck) and 300 µl EDTA (0.5M). The final volume was made with 15.34 ml sterile DDW and the solution was vortexed briefly and stored at room temperature. During the loading of DNA, the gel-loading buffer was diluted to 1× with the addition of 1× TBE.

- **Gel-Staining Dye**

A 10-mg pack of ethidium bromide (0.5 µg) was dissolved in 1 ml sterile triple DW. The solution was mixed and vortexed thoroughly. The solution was stored at room temperature and used for staining the DNA gel at a working concentration of 0.5 µg/ml (Sambrook et al. 2001).

Agarose Gel Electrophoresis

The extracted genomic DNA was analysed for its quality and integrity by running in 1% agarose gel in the 1× TBE buffer system. For gel casting, 1 g of agarose was added in 100 ml 1× TBE buffer solution. The agarose solution was warmed and allowed to melt resulting in a clear solution. Meanwhile, the gel-casting tray was sealed with adhesive tape and properly placed in a horizontal gel-casting table. When the agarose solution cooled down to about 60 °C, 4 µl of ethidium bromide was added, then poured carefully into the gel tray, immediately placing the comb, avoiding any air bubble. The gel was allowed to solidify for 45 min. After solidification, the comb and seal tape were carefully removed and the gel was immersed in the gel trough in the proper orientation with respect to the cathode and anode and immersed in sufficient 1× TBE buffer.

For electrophoresis of the DNA sample, 5 µl of DNA was mixed with 5 µl of the gel-loading dye on a parafilm paper and loaded onto each well. The gel was allowed to run at a constant voltage of 60 V (5V/cm distance between electrodes) for 1.2 hr till the tracking dye reached the middle of the gel. Photographs of the gel were taken using a Gel Doc system (BioRad, Hercules, CA, USA).

3.19 PCR Amplification of DNA Using ISSR Primers

A set of 15 inter-simple sequence repeat (ISSR; UBC, Vancouver, BC, Canada) primers were used for initial screening. Out of 15 ISSR primers screened, five primers yielded consistent and reproducible patterns of amplified products in all the plants. The details on the sequence of all the primers are given as follows:

UBC primer	Primer sequence (5'–3')
UBC 811	GAGAGAGAGAGAGAGAC
UBC 812	GAGAGAGAGAGAGAGAA
UBC 813	CTCTCTCTCTCTCTCTT
UBC 814	CTCTCTCTCTCTCTCTA
UBC 815	CTCTCTCTCTCTCTCTG
UBC 816	CACACACACACACACAT
UBC 817	CACACACACACACACAA
UBC 818	CACACACACACACACAG
UBC 819	GTGTGTGTGTGTGTGTA
UBC 820	GTGTGTGTGTGTGTGTC
UBC 821	GTGTGTGTGTGTGTGTT
UBC 822	TCTCTCTCTCTCTCTCA
UBC 823	TCTCTCTCTCTCTCTCC
UBC824	TCTCTCTCTCTCTCTCG
UBC 825	ACACACACACACACACT

3.19.1 Polymerase Chain Reaction Amplification

All the polymerase chain reaction (PCR) amplification reactions were performed in the Biometra PCR system.

The PCR reaction for the ISSR assay was carried out using the following amplification re-

agents (Fermentas, Genetix Biotech, Asia Pvt. Ltd. New Delhi, India):
a. 10× PCR buffer
b. 10 mM of deoxynucleotide triphosphates (dNTPs) mix
c. 25 mM $MgCl_2$ solution was supplied
d. 5 U/µl *Taq* DNA polymerase
e. 15 ng of ISSR primers (Genei, Banglore, India)

3.19.2 ISSR-PCR with Genomic DNA

PCR amplifications were carried out in a total volume of 30 µl containing 5 µl (20 ng) of genomic DNA. The reaction buffer consisted of 3 µl of 10× buffer, 0.75 µl $MgCl_2$ (25 mM), 0.75 µl dNTPs (10 mM each of deoxyadenosine triphosphate, dATP; deoxyguanosine triphosphate, dGTP; deoxythymidine triphosphate, dTTP and deoxycytidine triphosphate, dCTP), 1.5 µl primer, 0.15 µl *Taq* DNA polymerase and 18.85 µl water. DNA amplification program consisted of cycles starting with step one at 94 °C for 5 min followed by 30 s at the same temperature of 94 °C, 42 °C for 30 s, 72 °C for 1 min followed by 35 repeated cycles and a final extension at 72 °C for 5 min and finally at 4 °C till use.

3.19.3 Analysis of PCR Products by Agarose Gel Electrophoresis

The PCR-amplified products of *B. aegyptiaca* were resolved by electrophoresis on 1.5 % agarose gel in TBE buffer stained with 0.5 µg/ml of ethidium bromide solution for 3 h at 50V. DNA fingerprints were visualized under UV light and photographed using a gel-documenting system (BioRad, Hercules, USA). ISSR-PCR analysis using a primer was repeated thrice to evaluate the banding pattern of the DNA samples.

3.19.4 Data Scoring and Analysis

Only distinct, reproducible and well-resolved fragments ranging from 100 to 1,000 bp were considered in the analysis. These bands were scored as either present (1) or absent (0) for each of the ISSR markers within the ten plants. Electrophoretic DNA bands of low visual intensity that could not be readily distinguished as present or absent were considered ambiguous markers and were not scored. The size of the amplification products was estimated using a ladder DNA marker (New England Biolabs, 50 ng/µl, 1 kb).

The binary matrix was used to estimate genetic similarities using Jaccard's coefficient (Jaccard 1908) and the similarity matrix was subjected to the unweighted pair group method of arithmetic averages (UPGMA) clustering in order to construct the phenetic dendrograms. The reliability and robustness of the phenograms were tested by bootstrap analysis for 1,000 bootstraps for computing probabilities in terms of percentage for each node of the tree using the DARWin software (Perrier and Jacquemoud-Collet 2006).

References

Aebi H (1984) Catalse *in vitro* methods. Methods in Enzymology. Academic Press Inc 105:121–126

Bradford MM (1976) A rapid and sensitive method for quantitation of microgram quantities of protein utilizing the principles of protein-dye binding. Ann Biochem 72:248–254

Dhindsa PS, Plumb-Dhindsa P, Thorpe TA (1981) Leaf senescence: correlated with increased levels of membrane permeability and lipid peroxidation and decreased levels of superoxide dismutase and catalase. J Exp Bot 32:93–101

Doyle JJ, Doyle JL (1990) Isolation of plant DNA from fresh tissue. Focus 12:13–15

Foyer CH, Halliwell B (1976) The presence of glutathione reductase in chloroplast: a proposed role in ascorbic acid metabolism. Planta 133:21–25

Gamborg OL, Miller RA, Ojima K (1968) Nutrients requirement of suspension culture of soybean root cells. Exp Cell Res 50:151–158

Jaccard P (1908) Nouvelles rescherches sur la distribution florale. Bull Soc Vaud Sci Nat 44:223–270

Johansen DA (1940) Plant microtechnique. Mc Graw-Hill Book Co. Inc., New York

Lloyd G, McCown B (1980) Commercially feasible micropropagation of mountain laurel, *Kalmia latifolia*, by use of shoot tip culture. Int Plant Prop Soc Proc 30:421–427

Mackinney G (1941) Absorption of light by chlorophyll solution. J Biol Chem 140:315–322

Maclachan S, Zalick S (1963) Plastid structure, chlorophyll concentration and free amino acid composition of chlorophyll mutant barley. Can J Bot 41(7):1053–1062

Murashige T, Skoog F (1962) A revised medium for rapid growth and bioassays with tobacco tissue cultures. Physiol Plant 15:473–497

Nakano Y, Asada K (1981) Hydrogen peroxide is scavenged by ascorbate specific peroxidase in spinach chloroplasts. Plant Cell Physiol 22:867–880

Perrier X, Jacquemoud-Collet JP (2006) DARwin software, Version 5.0.158, http://darwin.cirad.fr/darwin

Phillips GC, Collins GB (1979) *In vitro* tissue culture of selected legumes and plant regeneration from callus of red clover. Crop Sci 19:59–64

Rao MV (1992) Cellular detoxifying mechanism determines age-dependent injury in tropical plants exposed to SO_2. J Plant Physiol 140:733–740

Sambrook J, Fritsch EJ, Maniatis T (2001) Molecular cloning: a laboratory manual. Cold Spring Harbor Laboratory Press, Cold Spring Harbor, New York

Results

Abstract

Balanites aegyptiaca (L.) Del., commonly known as Desert Date or Hingota is an evergreen spiny xerophytic tree. This plant can endure in a wide range of climatic conditions. This plant has got incredible importance and being used in treatment of a number of diseases since ages. The plant parts like roots, fruits contain diosgenin which can be used in pharmaceutical industry in production of oral contraceptives and steroids. Also the seed oil can be used as biofuel as combustion energy of oil is comparable to that of diesel. The present chapter deals with the observations obtained on enormous amount of studies undertaken in different aspects of plant tissue culture and molecular biology in *Balanites aegyptiaca* Del.

4.1 Direct Shoot Regeneration

4.1.1 Establishment of Aseptic Seedling

Seed germination was attempted to obtain young and surface sterile plant material for subsequent establishment of in vitro propagation protocols, necessary for the intended exploitation of the medicinal plant, *Balanites aegyptiaca*, through biotechnological means. Among the different strengths of Murashige and Skoog (MS) basal medium tested, full-strength MS medium was found superior resulting in 82% seed germination with the formation of healthy seedlings after 4 weeks (Fig. 4.1d), while a very low germination rate (20%) was recorded in ¼ MS medium (Table 4.1). The growing seedlings were then employed as a source of cotyledonary node (CN) explant and nodal and root explants which were subsequently used for in vitro propagation.

4.1.2 Regeneration from CN Explant Excised from 15-day-old Aseptic Seedlings

4.1.2.1 Effect of Cytokinins

MS medium devoid of plant growth regulator (PGR) failed to elicit regeneration even after 4 weeks; in contrast, the addition of cytokinins to the medium proved to be significantly useful in multiple shoot regeneration from CN explants. The data pertaining to the effect of cytokinins, viz., benzyladenine (BA), kinetin (Kn), and 2-isopentenyladenine (2-iP), on multiple shoot induction in a one-way factorial experiment are presented in Table 4.2. The explants grown on media supplemented with different concentrations of BA, Kn, and 2-iP (1.0, 2.5, 5.0, 10.0, 12.5, and 15.0 µM) initially responded with the enlargement and swelling of the nodes and started differentiating axillary buds directly within 1 week of incubation. The explants responded dif-

Fig. 4.1 a Selected candidate plus tree of *Balanites aegyptiaca* growing at Arid Forest Research Institute (AFRI) Campus, Jodhpur (Rajasthan, India). **b** and **c** Mature fruits and seeds of *B. aegyptiaca*. **d** Aseptic seedling on MS basal medium after 4 weeks of culture

Table 4.1 Effect of different strengths of MS medium on seed germination after 4 weeks of incubation

Strength of MS medium	Seed germination (%)
MS	82
½ MS	60
⅓ MS	40
¼ MS	20

MS Murashige and Skoog

ferently to different cytokinins but BA induced more shoots per explant compared to Kn and 2-iP. The range of percentage of shoot-regenerating explants was 28–80 %.

The explants cultured on a medium containing BA (1.0 µM) exhibited 49 % of regeneration frequency and produced 1.73 ± 0.26 shoots per explant with an average shoot length of 0.76 ± 0.03 cm. Shoot development was increased with an increase in the concentration of cytokinins up to a certain threshold, beyond which the frequency of shoot development was

4.1 Direct Shoot Regeneration

Table 4.2 Effect of cytokinins on multiple shoot induction from cotyledonary node explants derived from aseptic seedlings after 4 weeks of culture

Cytokinins (µM)			% Response	Mean no. of shoots/explants	Mean shoot length (cm)
BA	Kn	2-iP			
1.0	–	–	49	1.73±0.26^{gh}	0.76±0.03^{i}
2.5	–	–	60	3.16±0.37^{d}	1.00±0.05^{ghi}
5.0	–	–	65	4.60±0.35^{c}	2.00±0.00^{bcd}
10.0	–	–	70	6.00±0.05^{b}	1.73±0.17^{bcde}
12.5	–	–	80	8.00±0.05^{a}	2.90±0.57^{a}
15.0	–	–	50	2.53±0.26^{ef}	1.00±0.00^{ghi}
–	1.0	–	45	1.16±0.17^{hi}	0.73±0.21^{i}
–	2.5	–	52	2.03±0.13^{fg}	1.06±0.12^{fghi}
–	5.0	–	66	2.00±0.05^{fg}	1.43±0.24^{defgh}
–	10.0	–	68	3.00±0.05^{de}	1.80±0.17^{bcde}
–	12.5	–	75	4.06±0.12^{c}	2.30±0.10^{b}
–	15.0	–	51	2.10±0.46^{fg}	1.56±0.17^{cdefg}
–	–	1.0	30	1.70±0.15^{gh}	0.83±0.08^{hi}
–	–	2.5	44	2.06±0.12^{fg}	1.10±0.10^{fghi}
–	–	5.0	56	2.00±0.00^{fg}	1.20±0.20^{efghi}
–	–	10.0	58	2.26±0.12^{fg}	1.66±0.08^{cdef}
–	–	12.5	64	3.00±0.05^{de}	2.06±0.12^{bc}
–	–	15.0	28	1.00±0.00^{i}	1.00±0.00^{ghi}

BA benzyladenine, *Kn* kinetin, *2-iP* 2-isopentenyladenine
Values represent means±SE (standard error). Means followed by the same letter within columns are not significantly different ($p=0.05$) using Duncan's multiple range test

reduced. Among the three cytokinins tested, BA (12.5 µM) showed the highest shoot regeneration frequency (80%), maximum number (8.00±0.05) of shoots per explant, and maximum shoot length (2.90±0.57 cm) after 4 weeks of culture (Fig. 4.2a). At the same concentration, Kn and 2-iP produced 4.06±0.12 shoots per explant in 75% cultures and 3.00±0.05 shoots in 64% cultures, respectively (Table 4.2; Fig. 4.2b, c).

When Kn or 2-iP (1.0–15.0 µM) was used singly, low number of shoots were produced, and analysis of variance (ANOVA) revealed significant differences in parameters in comparison to BA (Table 4.2). Hence, explants grown on BA amended medium showed better growth and were found to be more responsive as compared to those grown on medium containing Kn and 2-iP. Based on the observations, BA at the concentration of 12.5 µM was selected to be optimum for obtaining maximum regeneration potential from CN explants.

4.1.2.2 Synergistic Effect of Cytokinins and Auxins

Using 12.5 µM as an optimum concentration of each cytokinin (BA, Kn, and 2-iP), three different auxins: α-naphthalene acetic acid (NAA), indole-3-acetic acid (IAA), and indole-3-butyric acid (IBA; 0.5, 1.0, 2.0, and 2.5 µM) were added to the MS medium to assess the synergistic effect of auxin and cytokinin on maximum shoot induction and proliferation from CN explant. The morphogenic responses are summarized in Tables 4.3, 4.4, and 4.5. Low concentration of auxins showed positive effect on multiple shoot induction, while higher concentrations resulted in callusing at the base of the explants; hence, reduction in the shoot number was recorded after 8 weeks.

Among the various concentrations of NAA with an optimal concentration of BA, Kn, and 2-iP, the highest frequency (90%) and superior number (14.73±0.29) of shoots per explant of length 3.96±0.08 cm was recorded on MS medium augmented with BA (12.5 µM) and NAA (1.0 µM) after 8 weeks (Table 4.3; Fig. 4.3a).

Fig. 4.2 a Production of multiple shoots in cotyledonary node explant on MS medium with 12.5 µM BA after 4 weeks of culture. **b** Multiple shoot initiation from a cotyledonary node explant in MS + 12.5 µM Kn after 4 weeks of culture. **c** Cotyledonary node explant showing multiple shoot formation on MS medium with 12.5 µM 2-iP after 4 weeks of culture. **d** Development of stunted multiple shoots from a cotyledonary node explants in MS + 5.0 µM thidiazuron (TDZ) after 4 weeks of culture

While, at the same concentration, Kn and NAA combination induced 11.96 ± 0.12 shoots per explant in 80% cultures (Table 4.4; Fig. 4.3b), 8.23 ± 0.14 shoots per explant were recorded in 2-iP + NAA combination in 77% cultures (Table 4.5).

Data analysis of Kn and/or 2-iP–IAA synergism revealed poor response as compared to BA–IAA conjunctions tried. Among the various concentrations of IBA along with optimal concentrations of BA, Kn, and 2-iP tested, MS medium fortified with BA (12.5 µM) and IBA (1.0 µM) produced only 9.36 ± 0.29 shoots per explant (Table 4.3) which was found fairly better than Kn–IBA and 2-iP–IBA combinations at the same concentration, respectively (Tables 4.4 and 4.5).

Table 4.3 Effect of auxins at different concentrations with an optimal concentration of BA (12.5 µM) in MS medium on shoot multiplication from cotyledonary node explants derived from aseptic seedlings after 8 weeks of culture

Auxins (µM)			% Response	Mean no. of shoots/explant	Mean shoot length (cm)
NAA	IAA	IBA			
0.5	–	–	70	9.00 ± 0.41^d	2.80 ± 0.15^c
1.0	–	–	90	14.73 ± 0.29^a	3.96 ± 0.08^a
2.0	–	–	82	12.00 ± 0.41^b	3.46 ± 0.26^b
2.5	–	–	66	5.10 ± 0.05^g	1.80 ± 0.11^e
–	0.5	–	64	4.50 ± 0.11^g	2.76 ± 0.14^c
–	1.0	–	71	11.00 ± 0.41^c	3.00 ± 0.05^c
–	2.0	–	68	7.16 ± 0.46^e	2.26 ± 0.12^d
–	2.5	–	59	3.16 ± 0.13^h	1.43 ± 0.24^{ef}
–	–	0.5	60	3.30 ± 0.23^h	1.20 ± 0.20^f
–	–	1.0	55	9.36 ± 0.29^d	2.50 ± 0.11^{cd}
–	–	2.0	59	6.00 ± 0.17^f	1.70 ± 0.15^e
–	–	2.5	50	1.73 ± 0.21^i	1.06 ± 0.12^f

NAA α-naphthalene acetic acid, *IAA* indole-3-acetic acid, *IBA* indole-3-butyric acid
Values represent means ± SE. Means followed by the same letter within columns are not significantly different ($p = 0.05$) using Duncan's multiple range test

Table 4.4 Effect of auxins at different concentrations with an optimal concentration of Kn (12.5 µM) in MS medium on shoot multiplication from cotyledonary node explants derived from aseptic seedlings after 8 weeks of culture

Auxins (µM)			% Response	Mean no. of shoots/explant	Mean shoot length (cm)
NAA	IAA	IBA			
0.5	–	–	66	4.13 ± 1.24^f	1.00 ± 0.00^e
1.0	–	–	80	11.96 ± 0.12^a	3.46 ± 0.26^a
2.0	–	–	75	8.26 ± 0.17^c	2.20 ± 0.11^{bc}
2.5	–	–	60	2.26 ± 0.17^g	1.70 ± 0.11^d
–	0.5	–	56	1.33 ± 0.14^h	1.06 ± 0.12^e
–	1.0	–	78	9.06 ± 0.12^b	2.26 ± 0.12^b
–	2.0	–	69	6.00 ± 0.17^d	1.73 ± 0.17^d
–	2.5	–	70	3.96 ± 0.08^f	1.80 ± 0.11^c
–	–	0.5	55	1.13 ± 0.08^h	0.76 ± 0.03^e
–	–	1.0	74	5.50 ± 0.26^e	1.80 ± 0.17^c
–	–	2.0	62	4.06 ± 0.12^f	1.20 ± 0.20^e
–	–	2.5	60	2.06 ± 0.12^g	1.00 ± 0.05^e

NAA α-naphthalene acetic acid, *IAA* indole-3-acetic acid, *IBA* indole-3-butyric acid
Values represent means ± SE. Means followed by the same letter within columns are not significantly different ($p = 0.05$) using Duncan's multiple range test

4.1.2.3 Effect of Thidiazuron and Subculturing

The CN explant established in a medium containing thidiazuron (TDZ) exhibited differentiation of the resident meristem, resulting in the formation of multiple shoots. The effect of different concentrations of TDZ on the shoot multiplication ability showed that the MS medium amended with TDZ (5.0 µM) gave a significantly higher number (5.26 ± 0.17) of shoots per explant with 70 % regeneration response than in the other treatments after 4 weeks of incubation (Table 4.6; Fig. 4.2d). Reduction in the parameters was noticed at higher concentrations beyond the optimal level. On increasing the concentration up to 15.0 µM, regeneration rate was decreased to 39 % forming 1.03 ± 0.11 shoot per explant accompanied by heavy basal callusing.

The regenerated shoot buds, induced on CN explant in TDZ-containing medium, when subcultured onto the same medium resulted in the formation of rosette of shoots which did not

Table 4.5 Effect of auxins at different concentrations with an optimal concentration of 2-iP (12.5 µM) in MS medium on shoot multiplication from cotyledonary node explants derived from aseptic seedlings after 8 weeks of culture

Auxins (µM)			% Response	Mean no. of shoots/explants	Mean shoot length (cm)
NAA	IAA	IBA			
0.5	–	–	60	3.00 ± 0.05^f	1.00 ± 0.00^{de}
1.0	–	–	77	8.23 ± 0.14^a	2.50 ± 0.11^a
2.0	–	–	65	6.00 ± 0.11^b	2.76 ± 0.18^a
2.5	–	–	62	4.16 ± 0.21^d	1.16 ± 0.17^{cde}
–	0.5	–	58	1.33 ± 0.14^h	0.73 ± 0.21^e
–	1.0	–	75	5.00 ± 0.11^c	2.06 ± 0.12^b
–	2.0	–	60	4.00 ± 0.11^c	1.43 ± 0.24^{cd}
–	2.5	–	53	2.06 ± 0.12^g	1.00 ± 0.00^{de}
–	–	0.5	50	1.00 ± 0.00^h	0.70 ± 0.11^e
–	–	1.0	65	3.46 ± 0.26^e	1.60 ± 0.11^e
–	–	2.0	45	2.06 ± 0.12^g	1.13 ± 0.08^{de}
–	–	2.5	40	1.13 ± 0.08^h	0.83 ± 0.08^e

NAA α-naphthalene acetic acid, *IAA* indole-3-acetic acid, *IBA* indole-3-butyric acid
Values represent means±SE. Means followed by the same letter within columns are not significantly different ($p=0.05$) using Duncan's multiple range test

elongate further. To surmount this problem, these shoot clusters were afterward transferred onto a fresh MS medium lacking TDZ at every 2 weeks. After the transfer, the shoots formed were healthy with well-developed leaves. The highest number (10.26 ± 0.12) of shoots per explant with shoot length of 4.00 ± 0.10 cm was achieved at the fourth subculture passage which became stabilized at the fifth passage (Fig. 4.4).

4.1.3 Regeneration from Nodal Explant Excised from 4-week old Aseptic Seedlings

4.1.3.1 Effect of Cytokinins

MS basal medium devoid of PGR did not support the induction of multiple shoots from the nodal explants. The number of shoots per responsive explant was significantly ($p=0.05$) affected by the type and concentration of various cytokinins used. The nodal segments responded by an initial enlargement of the dormant axillary buds followed by bud break within a week, and multiple shoot induction and proliferation were observed within 4 weeks of culture on cytokinin-containing media. The shoot bud differentiation was confirmed by histological study of nodal explants. Most shoot buds showed no visible connection with the original vascular tissue, although they appeared to have originated from the meristematic zone beneath the epidermis. It is possible that the initial shoots developed from preexisting meristems, as seen in histological slides (Fig. 4.22a, b, c). At 2.5 µM BA, 55% nodal explants produced 2.20 ± 0.41 shoots with a mean shoot length of 1.76 ± 0.31 cm after 4 weeks of culture. The frequency of axillary shoot proliferation and the number of shoots per explant increased with increasing concentration of BA, up to the optimal level such that BA at 12.5 µM showed the highest shoot regeneration frequency (80%) and the highest number (10.06 ± 0.86) of regenerated shoots per explant with an average shoot length of 4.56 ± 0.29 cm after 4 weeks of culture (Table 4.7; Fig. 4.5a). The frequency of shoot induction was relatively low and there were fewer shoots per explant when medium was supplemented with Kn (12.5 µM) or 2-iP (12.5 µM) instead of BA (12.5 µM; Fig. 4.5b, c). Furthermore, it was observed that the addition of higher concentrations in the medium (15.0 µM) of BA, Kn, or 2-iP resulted in callus growth at the basal end of the explants, which resulted in the inhibition of shoot number and shoot elongation.

4.1.3.2 Synergistic Effect of Cytokinins and Auxins

For further proliferation and elongation in multiple shoots induced from aseptic nodal explants,

4.1 Direct Shoot Regeneration

Fig. 4.3 **a** and **b** Healthy shoots formed vigorously from cotyledonary node explant on MS medium amended with BA (12.5 µM) + NAA (1.0 µM) and Kn (12.5 µM) + NAA (1.0 µM) after 8 weeks of culture, respectively

Table 4.6 Effect of TDZ on multiple shoot induction from cotyledonary node explants derived from aseptic seedlings after 4 weeks of culture

TDZ (µM)	% Response	Mean no. of shoots/explant	Mean shoot length (cm)
1.0	50	2.03 ± 0.08^c	1.70 ± 0.11^{ab}
2.5	61	3.30 ± 0.11^b	1.90 ± 0.05^a
5.0	70	5.26 ± 0.17^a	2.00 ± 0.11^a
10.0	68	4.00 ± 0.76^b	1.30 ± 0.15^{bc}
12.5	49	2.00 ± 0.00^c	1.03 ± 0.14^{cd}
15.0	39	1.03 ± 0.11^c	0.83 ± 0.20^d

TDZ thidiazuron
Values represent means ± SE. Means followed by the same letter within columns are not significantly different ($p = 0.05$) using Duncan's multiple range test

the optimal concentration (12.5 µM) of each cytokinin (BA, Kn, and 2-iP) was tested with different concentrations of different auxins, viz., NAA, IAA, and IBA (0.5, 1.0, 2.0, and 2.5 µM) and data obtained are depicted in Tables 4.8, 4.9, and 4.10. The regeneration frequencies and shoot production were significantly influenced by the cytokinin–auxin combinations, and higher multiplication rates were observed using MS medium amended with BA and NAA (Table 4.8). These treatments also resulted in the formation of more vigorous shoots with dark green, stiff, and shiny leaves. Initially, the low concentration of auxins (0.5 µM) with an optimal cytokinin in the medium enhanced the shoot production as compared to the single cytokinin treatment.

Moreover, among the BA–NAA conjunction tested, the best response was obtained on MS medium supplemented with BA (12.5 µM) and NAA (1.0 µM), resulting in the production of 18.00 ± 0.11 shoots per explant of shoot length 5.80 ± 0.23 cm after 8 weeks of incubation in 85 % cultures (Fig. 4.6a). At a higher concentration of NAA (2.5 µM), the number of shoots per explant reduced to 9.30 ± 0.35 after 8 weeks in 75 % cultures. However, the addition of Kn (12.5 µM) and NAA (1.0 µM) into the culture medium produced 13.00 ± 0.15 shoots with a shoot length of 4.53 ± 0.32 cm in 75 % cultures (Fig. 4.6b), and the supplementation of 2-iP (12.5 µM) and NAA (1.0 µM) produced 9.66 ± 0.88 shoots per explant with a shoot length of 3.73 ± 0.24 cm in 70 % cultures after 8 weeks of culture.

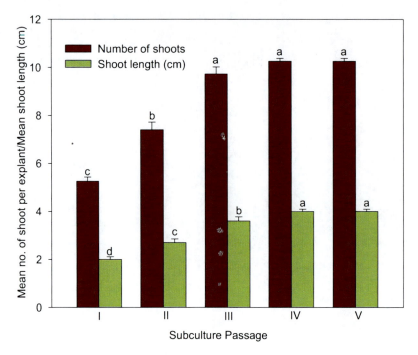

Fig. 4.4 Growth and development of shoots obtained from CN explants subcultured for different passages on MS medium without TDZ. Bars represent means ± SE. Bars denoted by the same letter within response variables are not significantly different ($p=0.05$) using Duncan's multiple range test

Table 4.7 Effect of cytokinins on multiple shoot induction from nodal explants derived from aseptic seedlings after 4 weeks of culture

Cytokinins (µM)			% Response	Mean no. of shoots/explant	Mean shoot length (cm)
BA	Kn	2-iP			
1.0	–	–	50	1.43±0.29^{fg}	1.20±0.20^{fgh}
2.5	–	–	55	2.20±0.41^{defg}	1.76±0.31^{efgh}
5.0	–	–	61	4.30±0.75^{cd}	3.03±0.43^{bcd}
10.0	–	–	66	8.00±1.15^{b}	4.16±0.61^{ab}
12.5	–	–	80	10.06±0.86^{a}	4.56±0.29^{a}
15.0	–	–	63	5.00±0.57^{c}	3.10±0.26^{bcd}
–	1.0	–	47	1.00±0.00^{g}	1.63±0.38^{fgh}
–	2.5	–	52	2.00±0.57^{efg}	2.36±0.27^{def}
–	5.0	–	59	3.63±0.71^{cdef}	2.90±0.20^{cde}
–	10.0	–	62	4.66±1.20^{c}	3.73±0.42^{abc}
–	12.5	–	70	5.00±1.15^{c}	4.00±0.86^{abc}
–	15.0	–	43	3.10±0.66^{cdefg}	2.40±0.28^{def}
–	–	1.0	37	1.00±0.39^{g}	0.80±0.11^{g}
–	–	2.5	46	1.53±0.39^{fg}	1.00±0.17^{gh}
–	–	5.0	49	2.20±0.61^{defg}	1.60±0.20^{fgh}
–	–	10.0	51	3.00±0.57^{cdefg}	2.03±0.26^{def}
–	–	12.5	65	4.20±0.41^{cde}	2.23±0.37^{def}
–	–	15.0	24	1.33±0.33^{g}	1.30±0.20^{fgh}

BA benzyladenine, *Kn* kinetin, *2-iP* 2-isopentenyladenine
Values represent means ± SE. Means followed by the same letter within columns are not significantly different ($p=0.05$) using Duncan's multiple range test

4.1 Direct Shoot Regeneration

Fig. 4.5 a Development of multiple shoots from seedling derived nodal explants on MS medium supplemented with 12.5 µM BA after 4 weeks of culture. **b** Shoot multiplication from seedling-derived nodal explant on MS medium containing 12.5 µM Kn after 4 weeks of culture. **c** The emergence of multiple shoots from the cultured seedling-derived nodal explants on MS medium augmented with 12.5 µM 2-iP after 4 weeks of culture. **d** Seedling-derived nodal explant cultured on MS medium containing 5.0 µM TDZ produced rosette of shoots after 4 weeks of culture

A similar situation was also recorded in the experiments where optimal concentrations of Kn and 2-iP were added in the medium augmented with IAA and IBA at different concentrations. The treatments resulted in the formation of multiple shoots, although with reduced efficiency (Tables 4.9 and 4.10).

4.1.3.3 Effect of TDZ and Subculturing

In the preliminary experiment, nodal explants obtained from aseptic seedlings were cultured for the multiple shoot induction on MS medium containing different concentrations of TDZ alone (Table 4.11). The explants failed to develop

Table 4.8 Effect of auxins at different concentrations with an optimal concentration of BA (12.5 μM) in MS medium on shoot multiplication from nodal explants derived from aseptic seedlings after 8 weeks of culture

Auxins (μM)			% Response	Mean no. of shoots/explant	Mean shoot length (cm)
NAA	IAA	IBA			
0.5	–	–	72	13.00±0.17^c	3.26±0.08^{de}
1.0	–	–	85	18.00±0.11^a	5.80±0.23^a
2.0	–	–	80	16.13±0.61^b	4.50±0.30^b
2.5	–	–	75	9.30±0.35^e	2.00±0.00^g
–	0.5	–	69	6.00±0.15^g	3.10±0.57^e
–	1.0	–	70	15.43±0.38^b	4.20±0.15^b
–	2.0	–	63	11.20±0.55^d	3.76±0.08^c
–	2.5	–	59	9.76±0.67^e	2.70±0.11^f
–	–	0.5	54	2.43±0.08^h	1.00±0.00^i
–	–	1.0	69	7.76±0.18^f	3.60±0.05^{cd}
–	–	2.0	50	4.96±0.24^g	2.13±0.08^g
–	–	2.5	44	2.03±0.08^h	1.60±0.05^h

NAA α-naphthalene acetic acid, *IAA* indole-3-acetic acid, *IBA* indole-3-butyric acid
Values represent means±SE. Means followed by the same letter within columns are not significantly different ($p=0.05$) using Duncan's multiple range test

Table 4.9 Effect of auxins at different concentrations with an optimal concentration of Kn (12.5L μM) in MS medium on shoot multiplication from nodal explants derived from aseptic seedlings after 8 weeks of culture

Auxins (μM)			% Response	Mean no. of shoots/explant	Mean shoot length (cm)
NAA	IAA	IBA			
0.5	–	–	66	8.43±1.40^{bc}	3.33±0.26^{bc}
1.0	–	–	75	13.00±0.15^a	4.53±0.32^a
2.0	–	–	70	10.00±0.15^b	4.20±0.30^{ab}
2.5	–	–	60	6.06±0.63^{cd}	2.06±0.23^{ab}
–	0.5	–	58	5.00±1.15^{defg}	2.93±0.23^{cd}
–	1.0	–	69	10.10±1.15^b	3.86±0.40^{ab}
–	2.0	–	62	8.66±0.66^{bc}	1.36±0.12^{fg}
–	2.5	–	50	2.56±0.43^{fg}	1.00±0.00^g
–	–	0.5	45	2.00±0.00^g	1.33±0.33^{fg}
–	–	1.0	58	7.66±1.76^{bc}	2.56±0.43^{cde}
–	–	2.0	49	5.33±0.88^{def}	1.96±0.26^{ef}
–	–	2.5	40	3.10±0.49^{efg}	2.20±0.30^{def}

NAA α-naphthalene acetic acid, *IAA* indole-3-acetic acid, *IBA* indole-3-butyric acid
Values represent means±SE. Means followed by the same letter within columns are not significantly different ($p=0.05$) using Duncan's multiple range test

shoots in a PGR-free MS medium even after 4 weeks of culture, while TDZ-containing medium promoted multiple shoot induction by the release of axillary buds from apical dominance within 2 weeks. However, modification of apical dominance was mainly dependent on TDZ concentration. Thus, the number of shoots per explant and frequency of shoot induction were mainly due to the concentration of TDZ used in the culture medium.

Of the various concentrations of TDZ tested, 5.0 μM TDZ proved to be optimum for multiple shoot induction. Increasing TDZ concentration above 5.0 μM had a negative effect on the frequency of shoot induction and number of shoots per explants. The highest frequency of shoot induction (80%), mean number (7.20±0.40) of shoots per nodal explant, and mean shoot length (3.90±0.17 cm) were obtained on MS medium supplemented with 5.0 μM TDZ after 4 weeks of culture (Table 4.11; Fig. 4.5d). High concentrations of TDZ resulted in callus formation at the base of the explants and hyperhydricity was induced in the regenerated shoots.

4.1 Direct Shoot Regeneration

Table 4.10 Effect of various auxins at different concentrations with an optimal concentration of 2-iP (12.5 μM) in MS medium on shoot multiplication from nodal explants derived from aseptic seedlings after 8 weeks of culture

Auxins (μM)			% Response	Mean no. of shoots/explant	Mean shoot length (cm)
NAA	IAA	IBA			
0.5	–	–	65	5.00 ± 1.15^{bcd}	1.00 ± 0.00^e
1.0	–	–	70	9.66 ± 0.88^a	3.73 ± 0.24^a
2.0	–	–	66	7.66 ± 2.33^{ab}	2.96 ± 0.06^b
2.5	–	–	52	4.66 ± 1.45^{bcde}	2.00 ± 0.05^{cd}
–	0.5	–	50	2.36 ± 0.32^{def}	1.13 ± 0.66^e
–	1.0	–	55	6.06 ± 0.63^{bc}	2.56 ± 0.43^{bc}
–	2.0	–	49	3.80 ± 0.30^{cdef}	2.00 ± 0.00^{cd}
–	2.5	–	35	1.76 ± 0.43^{ef}	1.63 ± 0.20^{de}
–	–	0.5	30	1.00 ± 0.17^f	1.00 ± 0.00^e
–	–	1.0	40	2.66 ± 0.66^{def}	2.00 ± 0.11^{cd}
–	–	2.0	29	1.76 ± 0.29^{ef}	1.53 ± 0.08^{de}
–	–	2.5	25	1.00 ± 0.00^f	1.03 ± 0.30^e

NAA α-naphthalene acetic acid, *IAA* indole-3-acetic acid, *IBA* indole-3-butyric acid
Values represent means ± SE. Means followed by the same letter within columns are not significantly different ($p = 0.05$) using Duncan's multiple range test

Table 4.11 Effect of TDZ on multiple shoot induction from nodal explants derived from aseptic seedlings after 4 weeks of culture

TDZ (μM)	% Response	Mean no. of shoots/explant	Mean shoot length (cm)
1.0	50	1.23 ± 0.14^d	2.00 ± 0.11^c
2.5	66	2.86 ± 0.17^c	2.70 ± 0.11^b
5.0	80	7.20 ± 0.40^a	3.90 ± 0.17^a
10.0	73	4.63 ± 0.20^b	3.03 ± 0.46^b
12.5	70	3.06 ± 0.12^c	1.86 ± 0.14^c
15.0	40	1.00 ± 0.00^d	1.00 ± 0.00^d

TDZ thidiazuron
Values represent means ± SE. Means followed by the same letter within columns are not significantly different ($p = 0.05$) using Duncan's multiple range test

The multiple shoots developed from the explants failed to elongate on the medium containing TDZ alone even after 6 weeks of culture. Moreover, hyperhydric shoots were formed which had malformed, wrinkled leaves and glossy, translucent appearance. The problem of shoot elongation and hyperhydricity was overcome by transferring shoot cultures on MS medium lacking TDZ. Subculturing of explants was carried out on a fresh medium at every 2 weeks. This process was repeated for five subcultures to study the effect of subculture on shoot multiplication. The number of shoots per explant increased up to four subculture passages and thereafter the number of shoots produced became stabilized (Fig. 4.7). Furthermore, the exclusion of TDZ from the medium significantly reduced the percentage of hyperhydricity and enhanced the multiplication rate with the development of healthy shoots.

4.1.3.4 Effect of Different Basal Media

Different basal media had different effects on shoot regeneration in nodal explant cultured on optimal concentration of cytokinin–auxin combination, i.e., BA (12.5 μM) and NAA (1.0 μM). The ANOVA indicated that the overall mean differences in the number of regenerated shoots per explant and shoot length due to different media were found to be highly significant. Among the different basal media tested, MS basal medium was favored the most for shoot regeneration, and it resulted in significantly greater number (18.00 ± 0.11) of shoots per explant followed by woody plant medium (WPM) and L_2 medium (Table 4.12). On B_5 medium, the regeneration frequency (53 %), number (5.33 ± 0.88) of shoots per explant, and shoot length (2.13 ± 0.08 cm) were very low. On the other hand, WPM and L_2 media had better effects on shoot regeneration; relatively higher frequencies (75 and 66 %, respectively) with good number of shoots were obtained after 8 weeks. It was also observed that on B_5 medium, the axillary shoots grew very

Fig. 4.6 Elongated shoots with well developed leaves regenerated from seedling derived nodal explants on MS medium with **a** 12.5 μM BA+1.0 μM NAA and **b** 12.5 μM Kn+1.0 μM NAA after 8 weeks of incubation, respectively

slowly and exhibited yellow color in comparison to other media.

4.1.3.5 Effect of Different Concentrations of Sucrose

No axillary shoots were induced on sucrose-free media even after 4 weeks of culture. The multiple shoot induction, from nodal explants cultured on MS medium containing BA (12.5 μM), NAA (1.0 μM), and different concentrations of sucrose, occurred within a week of incubation. On a medium containing sucrose, shoot regeneration frequencies and the number of shoots per explants increased with sucrose concentrations at 2–3%. Regeneration frequencies ranged from 78 to 85%, with increasing shoot number from 12.66 ± 0.88 to 18.00 ± 0.11 per explant. When sucrose concentration was further increased up to 5%, both regeneration frequencies and shoot number were inhibited, i.e., regeneration frequency was 65% with 4.00 ± 0.57 shoots per explant. The shoots that developed on media containing 1, 4, 5, and 6% sucrose were mostly greenish-yellow, whereas those developed with 2 and 3% were dark green and healthy throughout the culture. At higher concentrations (6%) of sucrose, callus formation occurred at the proximal end of the shoots. Thus, based on these results, the standard sucrose concentration of 3% was found to be the best in terms of regeneration frequency (85%), shoot number (18.00 ± 0.11) per explant, shoot length (5.80 ± 0.23 cm), and their subsequent growth after 8 weeks of culture (Table 4.13).

4.1.3.6 Effect of Different pH Levels

Our results showed that different pH levels 5.0–6.6 were broadly effective for shoot regeneration. Lower and higher pH levels showed low performance for the induction and proliferation of shoots. Furthermore, on the medium having pH 6.6 or below 5.4, the number of regenerated shoots was low. Moreover, on the medium pH value 5.0, regenerated shoots showed serious vitrification with the production of 7.66 ± 1.76 shoots per explants after 8 weeks of incubation.

4.1 Direct Shoot Regeneration

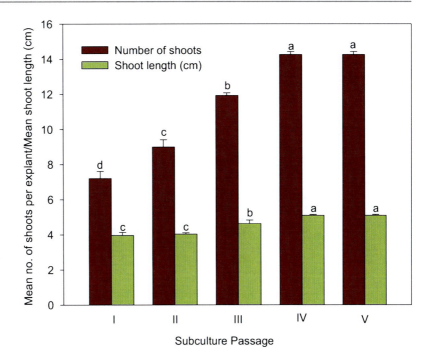

Fig. 4.7 Growth and development of shoots obtained from nodal explants excised from aseptic seedlings subcultured for different numbers of passages on MS medium without TDZ. Bars represent means±SE. Bars denoted by the same letter within response variables are not significantly different ($p=0.05$) using Duncan's multiple range test

Table 4.12 Effect of different media supplemented with BA (12.5 µM) and NAA (1.0 µM) on shoot induction and proliferation from nodal explants derived from aseptic seedlings after 8 weeks of culture

Media	% Response	Mean no. of shoots/explant	Mean shoot length (cm)
MS	85	18.00 ± 0.11^a	5.80 ± 0.23^a
WPM	75	10.00 ± 1.52^b	4.20 ± 0.15^b
L_2	66	7.66 ± 0.88^{bc}	3.60 ± 0.05^c
B_5	53	5.33 ± 0.88^c	2.13 ± 0.08^d

MS Murashige and Skoog, *WPM* woody plant medium
Values represent means±SE. Means followed by the same letter within columns are not significantly different ($p=0.05$) using Duncan's multiple range test

Table 4.13 Effect of sucrose on shoot induction and proliferation from nodal explants derived from aseptic seedlings after 8 weeks on MS medium supplemented with 12.5 µM BA and 1.0 µM NAA at pH 5.8

Concentrations (%)	% Response	Mean no. of shoots/explant	Mean shoot length (cm)
1	30	1.46 ± 0.14^e	0.76 ± 0.17^f
2	78	12.66 ± 0.88^b	4.43 ± 0.29^b
3	85	18.00 ± 0.11^a	5.80 ± 0.23^a
4	75	9.66 ± 0.88^c	3.10 ± 0.26^c
5	65	4.00 ± 0.57^d	2.13 ± 0.08^d
6	50	2.00 ± 0.05^e	1.46 ± 0.14^e

Values represent means±SE. Means followed by the same letter within columns are not significantly different ($p=0.05$) using Duncan's multiple range test

Among the different pH levels tested, pH 5.8 in MS medium containing BA (12.5 µM) and NAA (1.0 µM) was found to be the best in inducing maximum shoot number (18.00 ± 0.11) in 85% cultures, whereas a very poor response was recorded in the same medium at pH level 6.6 with the induction of only 1.00 ± 0.17 shoot per nodal explant in 40% cultures after 8 weeks (Table 4.14).

4.1.4 Regeneration from Root Explants Excised from 4-week-old Aseptic Seedlings

4.1.4.1 Effect of Cytokinins

Adventitious buds were formed from root explants in all the treatments except higher concentrations of each cytokinins (BA, Kn, and 2-iP) and statistically significant differences in

Table 4.14 Effect of initial medium pH on shoot induction and proliferation from nodal explants derived from aseptic seedlings on MS medium supplemented with 12.5 µM BA, 1.0 µM NAA and 3 % sucrose after 8 weeks of culture

pH levels	% Response	Mean no. of shoots/explant	Mean shoot length (cm)
5.0	70	7.66 ± 1.76^c	2.70 ± 0.11^c
5.4	73	11.20 ± 0.55^b	3.60 ± 0.05^b
5.8	85	18.00 ± 0.11^a	5.80 ± 0.23^a
6.2	50	2.43 ± 0.08^d	1.60 ± 0.05^d
6.6	40	1.00 ± 0.17^d	1.00 ± 0.00^e

Values represent means ± SE. Means followed by the same letter within columns are not significantly different ($p = 0.05$) using Duncan's multiple range test

the adventitious shoot regeneration capacity of root explants were noted in almost all the shoot induction media (Table 4.15). The adventitious bud formation was induced as a direct process without the formation of callus. Enlarged and developed protuberances (or nodular meristemoids) were initially observed in the middle as well as at the cut ends of the root explants within 2 weeks of culture (Fig. 4.8a, b). Subsequently, the protuberances differentiated into dark green adventitious buds which underwent normal growth and development (Fig. 4.8c, d). Furthermore, histology confirmed the direct organogenesis pathway showing shoot differentiation from root explants and maintaining vascular connections (Fig. 4.22d). Root explants inoculated on PGR-free medium yielded no shoot buds, while the frequency of bud formation on a medium containing cytokinins ranged from 18 to 54 %. Cytokinins at higher concentrations produced callus and did not show any morphogenesis even after 4 weeks of culture. Among the three cytokinins tested, 5.0 µM BA was found to be the most effective in induction of maximum number (4.50 ± 0.32) of shoots per explant and shoot length (5.86 ± 0.44 cm), whereas at the same concentration, Kn and 2-iP induced 3.46 ± 0.26 and 2.00 ± 0.05 shoots per explant, respectively, after 4 weeks of induction (Table 4.15; Fig. 4.9a, b, c). Moreover, the growth of the shoots remained arrested after 4 weeks of culture onto their respective media. The multiple shoots were then transferred on the media containing cytokinin–auxin combinations after 4 weeks for further growth.

4.1.4.2 Synergistic Effect of Cytokinins and Auxins

A synergistic influence of auxin and cytokinin was evident when combinations of optimal concentration of each cytokinin (BA, Kn, and 2-iP) with different concentrations of NAA, IAA, and IBA were tested during the study. The results are summarized in Tables 4.16, 4.17, and 4.18. Upon transfer of the buds on the shoot multiplication and elongation media, new adventitious buds started to multiply, while the preexisting adventitious shoots proliferated rapidly within 2 weeks followed by the formation of leaves and elongation of shoots in another 2 weeks of incubation. Moreover, the addition of NAA in the BA-containing medium enhanced the multiplication and elongation of shoots induced from the root explants and this was found to be most effective than the media containing IAA or IBA. Among various cytokinin–auxin combinations used, the highest shoot regeneration frequency (68 %), mean number of shoots per explant (7.20 ± 0.15), and mean shoot length (4.93 ± 0.03 cm) were recorded in MS medium containing a combination of 5.0 µM BA and 1.0 µM NAA after 8 weeks of culture (Table 4.16; Fig. 4.10a). Among different concentrations of IAA along with BA (5.0 µM), 1.0 µM IAA induced 3.13 ± 0.06 shoots per explants with a shoot length of 3.60 ± 0.15 cm in 63 % cultures after 8 weeks. Likewise, at the same concentration of BA–IBA combination, 2.53 ± 0.41 shoots per explants were recorded after 8 weeks in 56 % cultures (Table 4.16).

Similarly, among Kn–auxins combinations, maximum number (4.03 ± 0.18) of shoots per explant and the longest shoot (3.23 ± 0.14 cm) was recorded on a medium supplemented with Kn (5.0 µM) and NAA (1.0 µM) in 65 % cultures (Table 4.17; Fig. 4.10b), whereas in the midst of 2-iP and auxins combinations, the highest number (2.56 ± 0.29) of shoots per explant with an average shoot length (2.00 ± 0.05 cm) was observed on a medium supplemented with 2-iP (5.0 µM)

4.1 Direct Shoot Regeneration

Table 4.15 Effect of cytokinins on multiple shoot induction from root explant excised from aseptic seedlings after 4 weeks of culture

Cytokinins (µM)			% Response	Mean no. of shoots/explant	Mean shoot length (cm)
BA	Kn	2-iP			
1.0	–	–	22	2.20 ± 0.43^{bcd}	2.93 ± 1.03^{bc}
2.5	–	–	31	2.80 ± 0.54^{bc}	3.86 ± 0.08^{b}
5.0	–	–	54	4.50 ± 0.32^{a}	5.86 ± 0.44^{a}
10.0	–	–	–	+	–
12.5	–	–	–	+	–
15.0	–	–	–	+	–
–	1.0	–	23	2.00 ± 0.60^{cd}	1.56 ± 0.72^{c}
–	2.5	–	28	2.03 ± 0.60^{cd}	3.16 ± 0.44^{bc}
–	5.0	–	40	3.46 ± 0.26^{ab}	3.63 ± 0.84^{b}
–	10.0	–	–	+	–
–	12.5	–	–	+	–
–	15.0	–	–	+	–
–	–	1.0	18	1.30 ± 0.43^{d}	1.33 ± 0.44^{c}
–	–	2.5	20	1.43 ± 0.23^{cd}	2.30 ± 0.35^{bc}
–	–	5.0	37	2.00 ± 0.05^{cd}	2.70 ± 0.61^{bc}
–	–	10.0	–	+	–
–	–	12.5	–	+	–
–	–	15.0	–	+	–

BA benzyladenine, *Kn* kinetin, *2-iP* 2-isopentenyladenine
+ represents no response. Values represent means ± SE. Means followed by the same letter within columns are not significantly different ($p = 0.05$) using Duncan's multiple range test

Table 4.16 Effect of auxins at different concentrations with an optimal concentration of BA (5.0 µM) in MS medium on shoot multiplication from root explants excised from aseptic seedlings after 8 weeks of culture

Auxins (µM)			% Response	Mean no. of shoots/explant	Mean shoot length (cm)
NAA	IAA	IBA			
0.5	–	–	60	3.90 ± 0.05^{b}	4.30 ± 0.26^{b}
1.0	–	–	68	7.20 ± 0.15^{a}	4.93 ± 0.03^{a}
2.0	–	–	59	3.16 ± 0.21^{c}	4.00 ± 0.05^{bc}
2.5	–	–	56	3.03 ± 0.13^{c}	3.80 ± 0.26^{bcd}
–	0.5	–	61	2.86 ± 0.12^{cd}	3.20 ± 0.05^{def}
–	1.0	–	63	3.13 ± 0.06^{c}	3.60 ± 0.15^{cde}
–	2.0	–	50	2.20 ± 0.23^{e}	3.33 ± 0.14^{def}
–	2.5	–	49	2.10 ± 0.20^{e}	3.00 ± 0.05^{efg}
–	–	0.5	50	2.36 ± 0.27^{de}	2.53 ± 0.38^{ghi}
–	–	1.0	56	2.53 ± 0.41^{cde}	2.80 ± 0.05^{fgh}
–	–	2.0	45	2.00 ± 0.05^{e}	2.33 ± 0.08^{hi}
–	–	2.5	40	2.00 ± 0.15^{e}	2.10 ± 0.28^{i}

NAA α-naphthalene acetic acid, *IAA* indole-3-acetic acid, *IBA* indole-3-butyric acid
Values represent means ± SE. Means followed by the same letter within columns are not significantly different ($p = 0.05$) using Duncan's multiple range test

and NAA (1.0 µM) after 8 weeks of incubation (Table 4.18). However, a consistent decline in the percent regeneration, the number of shoots per explants, and shoot length was observed with an increase in the concentrations of auxins (2.0 µM, 2.5 µM) as it resulted in the formation of stunted shoots.

4.1.4.3 Effect of TDZ and Subculturing

The responding root explants swelled and turned light green, whereas the nonresponding explants

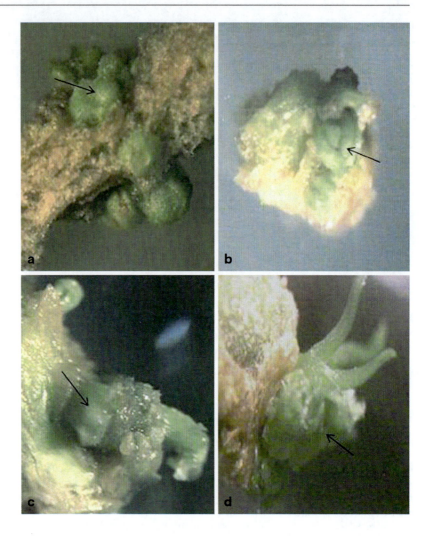

Fig. 4.8 a and **b** Emergence of direct shoot buds from root explant on MS medium augmented with 5.0 µM BA within 2 weeks of culture. **c** and **d** Advanced stages of culture showing differentiation of shoots

turned brown. Regeneration was evident after 2 weeks, with multiple clusters of dark green protuberances appearing from the surface and the cut ends of the explants cultured in TDZ media. Data generated from the experiment conducted with root explants to optimize the concentration of TDZ demonstrated that there was no bud formation in the absence of TDZ in the media, whereas statistical difference was observed between the different concentrations of TDZ on root explants in terms of overall average results (Table 4.19). The range of percentage of shoot-regenerating explants was 10–50% and the average number of shoot buds per explant varied significantly at different concentrations of TDZ (Table 4.19). The frequency of adventitious shoot induction and the number of shoots per explant increased with the increase in concentration of TDZ up to an optimum level. The greatest multiple shoot formation rate was obtained on MS medium supplemented with 2.5 µM TDZ. This enabled a 50% regeneration rate and produced an average of 3.03 ± 0.08 shoots more than 2.63 cm long per regenerating explant (Table 4.19; Fig. 4.9d). Increasing the TDZ concentration to 15.0 µM resulted in callus formation and no shoot morphogenesis was observed even after 4 weeks of incubation.

The multiple shoot buds obtained on various concentrations of TDZ showed vitrification and failed to elongate even after 4 weeks of culture.

Fig. 4.9 a Adventitious shoots produced directly from root explants after 4 weeks of in vitro culture on MS medium supplemented with 5.0 µM BA. **b** Direct adventitious shoots multiplied in root explant on MS medium containing 5.0 µM Kn after 4 weeks of culture. **c** Multiplication of adventitious shoots from root explants on MS medium augmented with 5.0 µM 2-iP after 4 weeks of culture. **d** Multiple shoot formation in root explant on TDZ (2.5 µM)-containing medium after 4 weeks of incubation

So, MS medium lacking TDZ was used in an effort to stimulate the shoot proliferation and elongation. After 4 weeks on TDZ-induced medium, adventitious buds were transferred to MS basal medium without PGRs and subcultured for five passages.

Repeated transfers in MS medium eliminated the deleterious effects of TDZ from the explants, thereby supporting differentiation of the buds to

Table 4.17 Effect of auxins at different concentrations with an optimal concentration of Kn (5.0 μM) in MS medium on shoot multiplication from root explants excised from aseptic seedlings after 8 weeks of culture

Auxins (μM)			% Response	Mean no. of shoots/explant	Mean shoot length (cm)
NAA	IAA	IBA			
0.5	–	–	61	3.03 ± 0.18^b	2.50 ± 0.34^{bc}
1.0	–	–	65	4.03 ± 0.18^a	3.23 ± 0.14^a
2.0	–	–	60	2.86 ± 0.12^{bc}	2.06 ± 0.12^{cd}
2.5	–	–	58	2.20 ± 0.23^{bcde}	1.73 ± 0.21^{def}
–	0.5	–	51	2.00 ± 0.15^{cde}	2.03 ± 0.03^{cd}
–	1.0	–	55	2.53 ± 0.32^{bcd}	2.80 ± 0.32^{ab}
–	2.0	–	57	2.20 ± 0.57^{bcde}	1.93 ± 0.31^{cde}
–	2.5	–	53	2.10 ± 0.37^{bcde}	1.56 ± 0.17^{def}
–	–	0.5	50	2.00 ± 0.50^{cde}	2.06 ± 0.21^{cd}
–	–	1.0	45	1.00 ± 0.00^f	2.13 ± 0.18^{cd}
–	–	2.0	40	1.80 ± 0.25^{def}	1.33 ± 0.18^{ef}
–	–	2.5	38	1.46 ± 0.08^{ef}	1.13 ± 0.03^f

NAA α-naphthalene acetic acid, *IAA* indole-3-acetic acid, *IBA* indole-3-butyric acid
Values represent means ± SE. Means followed by the same letter within columns are not significantly different ($p = 0.05$) using Duncan's multiple range test

Table 4.18 Effect of auxins at different concentrations with an optimal concentration of 2-iP (5.0 μM) in MS medium on shoot multiplication from root explants excised from aseptic seedlings after 8 weeks of culture

Auxins (μM)			% Response	Mean no. of shoots/explant	Mean shoot length (cm)
NAA	IAA	IBA			
0.5	–	–	46	2.00 ± 0.00^{ab}	1.80 ± 0.17^{ab}
1.0	–	–	50	2.56 ± 0.29^a	2.00 ± 0.05^a
2.0	–	–	36	1.80 ± 0.11^{bc}	1.56 ± 0.08^{abcd}
2.5	–	–	30	1.73 ± 0.21^{bcd}	1.03 ± 0.03^d
–	0.5	–	40	2.03 ± 0.08^{ab}	1.33 ± 0.14^{bcd}
–	1.0	–	45	2.10 ± 0.20^{ab}	1.70 ± 0.05^{abc}
–	2.0	–	38	1.60 ± 0.15^{bcde}	1.63 ± 0.14^{abcd}
–	2.5	–	30	1.23 ± 0.08^{cde}	1.40 ± 0.11^{abcd}
–	–	0.5	22	1.00 ± 0.00^e	1.06 ± 0.47^{cd}
–	–	1.0	25	1.13 ± 0.12^{cde}	1.26 ± 0.12^{bcd}
–	–	2.0	20	1.93 ± 0.27^{cd}	1.36 ± 0.24^{bcd}
–	–	2.5	19	1.06 ± 0.18^{de}	1.33 ± 0.53^{bcd}

NAA α-naphthalene acetic acid, *IAA* indole-3-acetic acid, *IBA* indole-3-butyric acid
Values represent means ± SE. Means followed by the same letter within columns are not significantly different ($p = 0.05$) using Duncan's multiple range test

healthy shoots. Although TDZ induced a good number of adventitious buds, all of these did not differentiate simultaneously. The excision of elongated shoots from the clusters possibly reduced the dominance exerted by the elongated shoots and hastened the elongation of remaining shoots. Furthermore, during the first to fourth passages, the number of shoots increased and thereafter it became stabilized at the fifth passage (Fig. 4.11). The regenerated shoots were healthy with well-developed leaves.

4.1.5 Induction of Multiple Shoots from Intact Seedlings

4.1.5.1 Effect of Gibberellic Acid on Seed Germination and Formation of Multiple Shoots

Seed germination took place within 2 weeks of inoculation. The seedling from the control (MS basal medium) developed one main shoot per seed with two or three nodes, whereas those from gibberellic acid (GA_3)-containing medium

4.1 Direct Shoot Regeneration

Fig. 4.10 Cultures showing an elongation and proliferation of healthy adventitious shoots induced from root explants after 8 weeks of incubation on MS medium containing **a** 5.0 μM BA + 1.0 μM NAA and **b** 5.0 μM Kn + 1.0 μM NAA, respectively

Table 4.19 Effect of TDZ on multiple shoot induction from root explant excised from aseptic seedlings after 4 weeks of culture

TDZ (μM)	% Response	Mean no. of shoots/explants	Mean shoot length (cm)
1.0	30	1.90 ± 0.05^b	1.53 ± 0.08^{bc}
2.5	50	3.03 ± 0.08^a	2.63 ± 0.31^a
5.0	40	2.23 ± 0.33^b	2.03 ± 0.08^b
10.0	10	1.00 ± 0.05^c	1.00 ± 0.00
12.5	–	+	–
15.0	–	+	–

TDZ thidiazuron
+ represents no response. Values represent means ± SE. Means followed by the same letter within columns are not significantly different ($p = 0.05$) using Duncan's multiple range test

developed—two to four shoots per seed, each containing one to two nodes after 4 weeks of culture. Elongation of the main shoot was inhibited, and axillary shoots were formed flanking the CN. However, no shoots were differentiated from the cotyledons or hypocotyl. The main root was long and thick without lateral roots. The green cotyledons remained unchanged in color and size compared to those on control. The bud clusters were formed on shoots which were not in contact with the medium (except for the roots). On medium containing 1.0 μM GA_3, 2.93 ± 0.58 shoots were produced per seedling after 4 weeks (Fig. 4.12a). Moreover, the number (4.23 ± 0.17) of shoots per seedling was the greatest with 2.0 μM GA_3 and the shoots attained a shoot length of 2.00 ± 0.28 cm in 4 weeks (Table 4.20, Fig. 4.12b). Although new shoot initials continued to form even after 4 weeks on GA_3-containing medium, the older shoots started browning and withered away.

4.1.5.2 Synergistic Effect of Cytokinins and Auxins

To overcome the initial low morphogenic response, the multiple shoots developed from seeds were excised and transferred to MS medium augmented with BA (12.5 μM) in synergy with NAA at different concentrations (0.5, 1.0, 2.0, and 5.0 μM). On medium containing BA (12.5 μM) and NAA (0.5 μM), 80% cultures showed the formation of 9.20 ± 0.62 shoots per

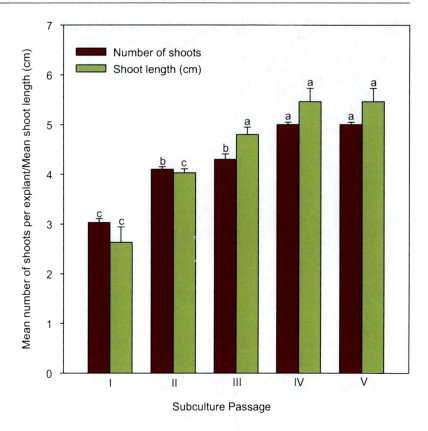

Fig. 4.11 Growth and development of shoots obtained from root explant excised from aseptic seedlings subcultured for different numbers of passages on MS medium without TDZ. Bars represent means ± SE. Bars denoted by the same letter within the response variables are not significantly different ($p=0.05$) using Duncan's multiple range test

Table 4.20 Effect of GA_3 at different concentrations in MS medium on germination and multiple shoot formation in seeds (intact seedlings) after 4 weeks of culture

Treatment (µM)	% Response	Mean no. of shoots/seed	Mean shoot length (cm)
0.0	82	1.00 ± 0.00[d]	0.53 ± 0.17[d]
0.5	29	1.73 ± 0.27[c]	1.00 ± 0.11[bcd]
1.0	60	2.93 ± 0.58[b]	1.50 ± 0.28[ab]
2.0	88	4.23 ± 0.17[a]	2.00 ± 0.28[a]
2.5	51	2.70 ± 0.15[bc]	1.23 ± 0.14[bc]
5.0	41	1.20 ± 0.28[c]	0.80 ± 0.05[cd]

Values represent means ± SE. Values followed by the different letters within columns are significantly different ($p = 0.05$) using Duncan's multiple range test

Table 4.21 Effect of NAA at different concentrations with an optimal concentration of BA (12.5 µM) on shoot multiplication of intact seedlings after 8 weeks of culture

NAA (µM)	% Response	Mean no. of shoots/ explant	Mean shoot length (cm)
0.5	80	9.20 ± 0.62[b]	5.26 ± 0.74[ab]
1.0	85	12.76 ± 0.14[a]	6.53 ± 0.37[a]
2.0	75	6.56 ± 0.65[c]	3.90 ± 0.17[b]
2.5	60	4.60 ± 0.34[d]	4.16 ± 0.76[b]

NAA α-naphthalene acetic acid

Values represent means ± SE. Values followed by the different letters within columns are significantly different ($p = 0.05$) using Duncan's multiple range test

culture after 8 weeks (Fig. 4.13a). These BA–NAA conjunctions accentuated the morphogenic response and facilitated the growth of axillary shoots significantly. Among all concentrations of the BA–NAA combination used, the maximum number (12.7 ± 0.14) of shoots and shoot length (6.53 ± 0.37 cm) per culture were obtained at 12.5 µM BA with 1.0 µM NAA after 8 weeks of culture (Table 4.21; Fig. 4.13b).

4.1.5.3 Effect of TDZ and Subculturing

The effect of TDZ at different concentrations was also tested to enhance the multiplication rate of the multiple shoots obtained from seeds. Among different concentrations used, TDZ (5.0 µM) produced maximum number of shoots per seed (4.50 ± 0.28) with a shoot length of 2.86 ± 0.12 cm after 4 weeks, whereas at higher concentrations (10.0–15.0 µM) the number of shoots declined (Table 4.22). However, the shoot multiplication

4.1 Direct Shoot Regeneration

Fig. 4.12 a Formation of multiple shoots from cultured intact seedling on MS medium containing 1.0 µM GA_3 after 4 weeks of culture. **b** Multiple shoots with expanded leaves developing from intact seedling on GA_3 (2.0 µM)-containing medium after 4 weeks of culture initiation

rate was increased when the shoot clusters were transferred to TDZ-free MS medium and subcultured for five passages. At the third subculture, the shoot number reached 5.80 ± 0.30 with an average length of 4.00 ± 0.05 cm. However, the maximum number (7.33 ± 0.38) of shoots was recorded at the fourth subculture passage and thereafter the number became stabilized at the fifth passage (Fig. 4.14). The regenerated shoots obtained were healthy with well-developed leaves.

4.1.6 Regeneration from Mature Nodal Explants Excised from 10-year-old Candidate Plus Tree

4.1.6.1 Effect of Cytokinins

The nodal explants failed to respond morphogenetically to a growth regulator-free MS medium.

Fig. 4.13 a and **b** Prolific elongation of multiple shoots excised from intact seedling on MS medium containing BA (12.5 µM) and NAA (1.0 µM) after 8 weeks of culture

Table 4.22 Effect of TDZ on multiple shoots excised from intact seedlings after 4 weeks of culture

TDZ (µM)	% Response	Mean no. of shoots/ explant	Mean shoot length (cm)
1.0	20	2.30 ± 0.11^c	1.76 ± 0.14^d
2.5	29	2.90 ± 0.20^b	2.23 ± 0.24^{bc}
5.0	67	4.50 ± 0.28^a	2.86 ± 0.12^a
10.0	30	2.26 ± 0.14^c	2.56 ± 0.17^{ab}
12.5	18	2.16 ± 0.16^c	1.23 ± 0.14^d
15.0	15	2.00 ± 0.00^c	1.16 ± 0.27^d

TDZ thidiazuron
Values represent means±SE. Means followed by the same letter within columns are not significantly different ($p = 0.05$) using Duncan's multiple range test

In contrast, on the media containing cytokinins, an enlargement and subsequent break of axillary buds were observed in all media tested within 2 weeks of explant inoculation. The percentage response of explants for shoot induction, shoot number, and shoot length varied according to the type and concentration of cytokinins used (Table 4.23). The dormant axillary buds of all explants sprouted and then developed to shoots with two to four pairs of leaves at 4 weeks of culture. The nodal explants cultured on cytokinin-containing media showed their first response by initial enlargement of the nodes with axillary buds followed by bud break after 2 weeks of inoculation. In another 2 weeks, the shoot buds elongated and developed into healthy shoots. All regenerated shoots were free from any basal callusing at their proximal ends. Among the three cytokinins tested, BA (12.5 µM) was found to be optimum for inducing maximum regeneration frequency (67%), number of shoots (5.06±0.12), and shoot length (3.73±0.06 cm) than Kn or 2-iP (Table 4.23; Fig. 4.15a). Moreover, increasing the concentration of cytokinins beyond the optimal level decreased the percentage of responding cultures as well as the number of shoots. The addition of Kn and 2-iP significantly reduced the average number of shoots at all the concentrations tested. On the medium containing Kn (12.5 µM), 4.26±0.06 shoots per explants were recorded after 4 weeks of culture initiation (Fig. 4.15b). Similarly, at the same concentration, 2-iP induced only 2.10±0.46 shoots per explant after 4 weeks (Fig. 4.15c). This way, of the three cytokinins tested, BA was found to be more effective than Kn and 2-iP in the induction of shoots from mature nodal explants.

4.1.6.2 Synergistic Effect of Cytokinins and Auxins

The synergistic influence of cytokinin with auxin on nodal explants was evident when a combination of optimal concentration (12.5 µM) of each cytokinin (BA, Kn, and 2-iP) with different concentrations of NAA, IAA, and IBA (0.5, 1.0, 2.0, and 2.5 µM) was tested. The addition of NAA at 1.0 µM concentration significantly enhanced the shoot regeneration capacity with the production of a good number of multiple shoots in MS medium supplemented with optimal concentration of each cytokinin (Table 4.24). Among all the BA–NAA combinations, the maximum number of shoots (7.70±0.11) per explant with an average shoot length of 4.40±0.11 cm was obtained at 12.5 µM BA with 1.0 µM NAA in 75% culture after 8 weeks (Fig. 4.16a). On increasing the concentration of NAA up to 2.5 µM, the basal callusing was observed at the cut end of the nodal explants; thus, the regeneration frequency and the number of shoots per explant were decreased. On the other hand, MS medium amended with BA and IAA and BA and IBA at various concentrations showed varying degrees of regeneration response in 67 and 63% cultures, respectively (Table 4.24).

Furthermore, a significantly low number of shoots (6.80±0.37) were produced in MS medium augmented with Kn (12.5 µM)±NAA (1.0 µM) combination in 65% cultures (Fig. 4.16b), and a very low number of shoots (3.00±0.05) were recorded on 2-iP (12.5 µM)±NAA (1.0 µM) combination in 50% cultures after 8 weeks (Tables 4.25 and 4.26). Among the three auxins tested in combination with optimal concentration of each cytokinin, NAA at 1.0 µM concentration gave the maximum response as compared to IAA and IBA.

4.1.6.3 Effect of TDZ and Subculturing

When mature nodal explants were cultured on MS medium amended with 5.0 µM TDZ, 60%

4.1 Direct Shoot Regeneration

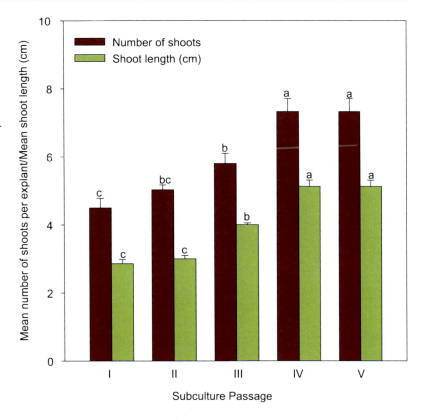

Fig. 4.14 Growth and development of shoots obtained from intact seedlings subcultured for different numbers of passages on MS medium without TDZ. Bars represent means ± SE. Bars denoted by the same letter within response variables are not significantly different ($p=0.05$) using Duncan's multiple range test

Table 4.23 Effect of cytokinins on multiple shoot induction from mature nodal explants after 4 weeks of culture

Cytokinins (µM)			% Response	Mean no. of shoots/explants	Mean shoot length (cm)
BA	Kn	2-iP			
1.0	–	–	40	1.23 ± 0.14^{gh}	1.16 ± 0.17^{ij}
2.5	–	–	55	1.90 ± 0.05^{def}	2.26 ± 0.17^{gh}
5.0	–	–	60	3.23 ± 0.17^{c}	3.06 ± 0.12^{bcd}
10.0	–	–	64	4.06 ± 0.12^{b}	2.66 ± 0.18^{def}
12.5	–	–	67	5.06 ± 0.12^{a}	3.73 ± 0.06^{a}
15.0	–	–	59	3.96 ± 0.17^{b}	1.33 ± 0.14^{ij}
–	1.0	–	42	1.03 ± 0.08^{h}	1.86 ± 0.03^{hi}
–	2.5	–	47	1.43 ± 0.24^{fgh}	3.16 ± 0.13^{bc}
–	5.0	–	52	2.30 ± 0.10^{d}	3.46 ± 0.26^{ab}
–	10.0	–	57	3.30 ± 0.17^{c}	2.76 ± 0.14^{bcd}
–	12.5	–	63	4.26 ± 0.06^{b}	2.50 ± 0.11^{fg}
–	15.0	–	54	2.40 ± 0.40^{d}	1.33 ± 0.21^{ij}
–	–	1.0	38	1.00 ± 0.00^{h}	0.83 ± 0.08^{j}
–	–	2.5	39	1.10 ± 0.05^{h}	1.20 ± 0.20^{ij}
–	–	5.0	55	1.56 ± 0.17^{efgh}	1.43 ± 0.24^{hi}
–	–	10.0	35	2.06 ± 0.12^{de}	1.56 ± 0.17^{hi}
–	–	12.5	30	2.10 ± 0.46^{de}	1.56 ± 0.06^{hi}
–	–	15.0	25	1.70 ± 0.15^{efg}	1.06 ± 0.12^{ij}

BA benzyladenine, *Kn* kinetin, *2-iP* 2-isopentenyladenine
Values represent means ± SE. Means followed by the same letter within columns are not significantly different ($p=0.05$) using Duncan's multiple range test

Fig. 4.15 a Sprouted axillary shoots from bud breaking in mature nodal explants on MS medium supplemented with 12.5 µM BA after 4 weeks of culture initiation. **b** Induction of multiple axillary shoots from mature nodal explant on MS medium containing 12.5 µM Kn after 4 weeks of culture. **c** Formation of axillary shoots on MS medium augmented with 12.5 µM 2-iP in mature nodal explant after 4 weeks of induction. **d** Multiplication of axillary shoots from mature nodal explant on TDZ (5.0 µM)-supplemented medium after 4 weeks of culture

of explants exhibited maximum bud break with 3.90 ± 0.11 shoots per explant and average shoot length (2.70 ± 0.11 cm) after 4 weeks of culture (Table 4.27; Fig. 4.15d). At higher concentrations (12.5–15.0 µM) of TDZ, frequency of bud break dropped dramatically with callus formation along the base of explants producing short shoots, while at lower TDZ concentrations (1.0–5.0 µM), frequency of bud break remained satisfactory. On MS medium supplemented with 5.0 µM TDZ, shoots exhibited hyperhydration and fasciation and showed declined growth when incubated for more than 4 weeks. So, to overcome the deleterious effect of TDZ, multiple shoots were subcultured on PGR-free MS medium for five passages after every 2 weeks. As a result, an increasing trend of shoot multiplication rate was observed during each passage. On the PGR-free MS medium,

Table 4.24 Effect of auxins at different concentrations with an optimal concentration of BA (12.5 µM) in MS medium on shoot multiplication from mature nodal explants after 8 weeks of culture

Auxins (µM)			% Response	Mean no. of shoots/explant	Mean shoot length (cm)
NAA	IAA	IBA			
0.5	–	–	68	5.63 ± 0.20^b	4.00 ± 0.05^b
1.0	–	–	75	7.70 ± 0.11^a	4.40 ± 0.11^a
2.0	–	–	72	4.46 ± 0.21^d	3.43 ± 0.08^c
2.5	–	–	70	3.30 ± 0.15^f	2.93 ± 0.20^{de}
–	0.5	–	56	3.16 ± 0.21^f	2.13 ± 0.20^f
–	1.0	–	67	5.03 ± 0.08^c	3.00 ± 0.11^d
–	2.0	–	60	3.90 ± 0.05^e	2.56 ± 0.08^e
–	2.5	–	57	2.60 ± 0.17^{gh}	2.03 ± 0.08^f
–	–	0.5	53	2.23 ± 0.17^{hi}	1.40 ± 0.05^h
–	–	1.0	63	3.00 ± 0.05^{fg}	1.96 ± 0.12^{fg}
–	–	2.0	50	2.30 ± 0.10^{hi}	1.60 ± 0.17^{gh}
–	–	2.5	45	2.00 ± 0.05^i	1.33 ± 0.14^h

NAA α-naphthalene acetic acid, *IAA* indole-3-acetic acid, *IBA* indole-3-butyric acid
Values represent means ± SE. Means followed by the same letter within columns are not significantly different ($p = 0.05$) using Duncan's multiple range test

Table 4.25 Effect of various auxins at different concentrations with an optimal concentration of Kn (12.5 µM) in MS medium on shoot multiplication from mature nodal explants after 8 weeks of culture

Auxins (µM)			% Response	Mean no. of shoots/explant	Mean shoot length (cm)
NAA	IAA	IBA			
0.5	–	–	50	4.20 ± 0.20^b	3.20 ± 0.05^{bc}
1.0	–	–	65	6.80 ± 0.37^a	4.10 ± 0.05^a
2.0	–	–	60	4.06 ± 0.12^b	3.56 ± 0.06^b
2.5	–	–	55	3.10 ± 0.10^c	3.00 ± 0.05^c
–	0.5	–	45	2.56 ± 0.24^{de}	2.33 ± 0.08^{def}
–	1.0	–	49	2.90 ± 0.05^{cd}	2.76 ± 0.17^{cd}
–	2.0	–	39	2.13 ± 0.03^{ef}	2.50 ± 0.11^d
–	2.5	–	40	2.06 ± 0.08^{ef}	2.26 ± 0.34^{efg}
–	–	0.5	32	1.60 ± 0.17^{fg}	2.00 ± 0.15^{fg}
–	–	1.0	36	1.96 ± 0.12^f	1.93 ± 0.08^{fg}
–	–	2.0	34	1.33 ± 0.14^g	1.80 ± 0.11^g
–	–	2.5	29	1.13 ± 0.12^g	1.80 ± 0.17^g

NAA α-naphthalene acetic acid, *IAA* indole-3-acetic acid, *IBA* indole-3-butyric acid
Values represent means ± SE. Means followed by the same letter within columns are not significantly different ($p = 0.05$) using Duncan's multiple range test

Fig. 4.16 a Proliferation and elongation of vigorous shoots induced from mature nodal explant on MS medium containing a combination of BA (12.5 µM) and NAA (1.0 µM) after 8 weeks of incubation. **b** Production of axillary shoots obtained from mature nodal explants on MS medium augmented with kinetin (12.5 µM) and NAA (1.0 µM) after 8 weeks of culture

Table 4.26 Effect of various auxins at different concentrations with an optimal concentration of 2-iP (12.5 μM) in MS medium on shoot multiplication from mature nodal explants after 8 weeks of culture

Auxins (μM)			% Response	Mean no. of shoots/explant	Mean shoot length (cm)
NAA	IAA	IBA			
0.5	–	–	45	2.36 ± 0.37[abc]	2.10 ± 0.37[abc]
1.0	–	–	50	3.00 ± 0.05[a]	2.56 ± 0.24[a]
2.0	–	–	40	2.73 ± 0.14[ab]	2.36 ± 0.29[ab]
2.5	–	–	39	2.43 ± 0.12[abc]	2.00 ± 0.11[abc]
–	0.5	–	35	2.10 ± 0.28[bcd]	1.96 ± 0.12[abc]
–	1.0	–	38	2.50 ± 0.23[abc]	2.06 ± 0.08[abc]
–	2.0	–	35	2.00 ± 0.15[cd]	1.80 ± 0.17[bc]
–	2.5	–	28	1.90 ± 0.11[cd]	1.96 ± 0.06[abc]
–	–	0.5	22	1.60 ± 0.17[de]	1.85 ± 0.09[bc]
–	–	1.0	25	2.00 ± 0.15[cd]	2.00 ± 0.10[abc]
–	–	2.0	20	1.93 ± 0.27[cd]	1.73 ± 0.20[bc]
–	–	2.5	20	1.23 ± 0.17[e]	1.63 ± 0.17[c]

NAA α-naphthalene acetic acid, *IAA* indole-3-acetic acid, *IBA* indole-3-butyric acid
Values represent means ± SE. Means followed by the same letter within columns are not significantly different ($p = 0.05$) using Duncan's multiple range test

Table 4.27 Effect of TDZ on multiple shoot induction from mature nodal explants after 4 weeks of culture

TDZ (μM)	% Response	Mean no. of shoots/explant	Mean shoot length (cm)
1.0	49	2.13 ± 0.08[c]	1.00 ± 0.00[e]
2.5	58	3.03 ± 0.08[b]	1.33 ± 0.14[d]
5.0	60	3.90 ± 0.11[a]	2.70 ± 0.11[a]
10.0	55	3.10 ± 0.11[b]	2.26 ± 0.06[b]
12.5	50	2.00 ± 0.00[c]	1.96 ± 0.03[c]
15.0	45	1.03 ± 0.06[d]	1.20 ± 0.05[de]

TDZ thidiazuron
Values represent means ± SE. Means followed by the same letter within columns are not significantly different ($p = 0.05$) using Duncan's multiple range test

8.03 ± 0.08 shoots differentiated per explant which was found to be the maximum at the fourth passage (Fig. 4.17). Furthermore, the regenerated shoots were healthy and the multiplication rate became stabilized at the fifth subculture passage.

4.1.7 Rooting of Microshoots

4.1.7.1 In Vitro Rooting

Shoots regenerated from nodal explants failed to produce roots on full-strength MS basal medium. Only one root was formed on half-strength MS basal medium after 4 weeks (Fig. 4.18a). The addition of auxin in half-strength MS medium was essential for root induction. So, to promote the development of the root system, 4–5 cm long shoots were transferred on MS medium augmented with IBA, NAA, and IAA alone at different concentrations (0.1, 0.5, 1.0, 2.0, and 5.0 μM; Table 4.28). The development of roots took place within 2 weeks of incubation without an intervening callus phase on all the auxin-containing rooting media. The augmentation of IBA to half-strength MS medium enhanced rooting significantly with the formation of healthy and longer roots. On the medium containing NAA and IAA singly, roots were thick and stunted. The highest rooting percentages were obtained in half-strength MS medium with 1.0 μM IBA (80%), followed by NAA (73%) at 1.0 μM. The addition of IAA at either lower or higher concentrations than 1.0 μM resulted in lower percentages of rooting as compared to those obtained with NAA and IBA. Data expressed as the number of roots per shoot and root length supported the trend established for the rooting experiment. On lower concentrations of IBA (0.1–0.5 μM), the number of roots per shoot was considerably less, i.e., 3.33 ± 0.29, and 3.38 ± 0.17 roots per shoot was recorded after 4 weeks of culture. Also, at a higher (5.0 μM) concentration of IBA, the number (2.60 ± 0.49) of roots and length (2.83 ± 0.17 cm) reduced. Similar

4.1 Direct Shoot Regeneration

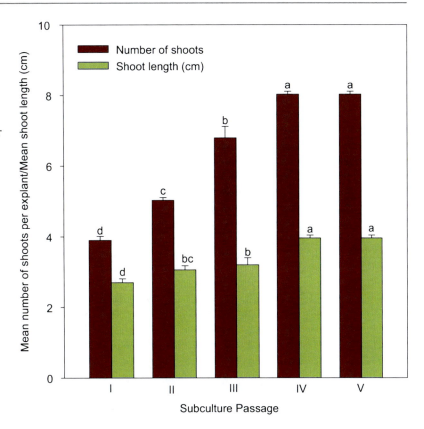

Fig. 4.17 Growth and development of shoots obtained from intact seedlings subcultured for different numbers of passages on MS medium without TDZ. Bars represent means ± SE. Bars denoted by the same letter within response variables are not significantly different ($p = 0.05$) using Duncan's multiple range test

Table 4.28 Influence of different auxins in ½ MS medium on in vitro rooting of the microshoots after 4 weeks of culture

Auxins (µM)			% Rooting	Mean no. of roots/shoot	Mean root length (cm)
IBA	NAA	IAA			
0.1	–	–	53	3.33 ± 0.29^{cde}	1.30 ± 0.32^{hij}
0.5	–	–	62	3.83 ± 0.17^{cd}	2.36 ± 0.29^{cdef}
1.0	–	–	80	8.10 ± 0.60^{a}	4.30 ± 0.37^{a}
2.0	–	–	70	5.63 ± 0.37^{b}	3.16 ± 0.17^{bc}
5.0	–	–	61	2.60 ± 0.49^{defg}	2.83 ± 0.17^{bcde}
–	0.1	–	45	2.60 ± 0.49^{defg}	1.10 ± 0.25^{ij}
–	0.5	–	60	2.80 ± 0.28^{def}	2.06 ± 0.23^{defgh}
–	1.0	–	73	4.30 ± 0.62^{c}	3.56 ± 0.23^{b}
–	2.0	–	64	2.26 ± 0.49^{efgh}	2.46 ± 0.26^{cde}
–	5.0	–	54	1.86 ± 0.08^{fghi}	1.83 ± 0.37^{efgh}
–	–	0.1	44	1.46 ± 0.27^{ghi}	0.90 ± 0.20^{j}
–	–	0.5	56	1.76 ± 0.35^{fghi}	1.56 ± 0.17^{ghij}
–	–	1.0	68	4.33 ± 0.33^{bc}	3.00 ± 0.05^{bc}
–	–	2.0	62	1.26 ± 0.23^{hi}	2.13 ± 0.08^{defg}
–	–	5.0	49	0.95 ± 0.05^{i}	1.50 ± 0.28^{ghi}

Values represent means ± SE. Values followed by different letters within columns are significantly different ($p = 0.05$) using Duncan's multiple range test

responses were also recorded for lower and higher concentrations of NAA and IAA (Table 4.28).

Nevertheless, the best rhizogenic response was recorded at 1.0 µM IBA where an average of 8.10 ± 0.60 roots of root length 4.3 ± 0.37 cm was recorded after 4 weeks of culture (Table 4.28; Fig. 4.18b, c). On this treatment, upper shoot growth was manifested in new leaf expansion and stem elongation. The roots were moderately thin which help in establishing the plantlets in the soil.

4.1.7.2 Ex Vitro Rooting

Rooting was also carried out by ex vitro method as described in Chap. 3. With the treatment given in 100 and 150 µM IBA solution, the roots were very thin and weak. Furthermore, at higher concentrations of IBA (250 and 300 µM), basal callusing was observed and no root formation took place even after 4 weeks of transfer. However, the best results were recorded at IBA (200 µM) as it gave the maximum number (4.03 ± 0.15) of roots

Fig. 4.18 **a** Induction of single root in microshoot on half-strength MS medium after 4 weeks of incubation. **b** and **c** In vitro rooted shoots cultured on half-strength MS medium with 2.0 µM IBA after 4 weeks. **d** Ex vitro rooted plantlet following pulse treatment with 200 µM IBA after 4 weeks of transfer

per shoot and root length (3.13 ± 0.18 cm) after 4 weeks of transplantation (Table 4.29; Fig. 4.18d). The roots produced with this treatment were healthy, strong, and fared well using single-pulse treatment of IBA together with plantlet transplantation in soilrite which eliminated the additional in vitro rooting step.

4.1 Direct Shoot Regeneration

Table 4.29 Effect of different concentrations of IBA on ex vitro rooting of microshoots after 4 weeks of transfer

IBA (µM)	% Rooting	Mean no. of roots/shoot	Mean root length (cm)
100	30	1.33 ± 0.38^b	1.53 ± 0.26^b
150	50	2.00 ± 0.20^b	1.90 ± 0.05^b
200	65	4.03 ± 0.15^a	3.13 ± 0.18^a
250	–	+	–
300	–	+	–

IBA indole-3-butyric acid
+ denotes callus formation. Values represent means ± SE. Values followed by different letters within columns are significantly different ($p = 0.05$) using Duncan's multiple range test

Table 4.30 Influence of different planting substrates for hardening off in vitro raised plantlets

Substratum	% survival rate in growth chamber after 4 weeks	% survival rate in the field conditions after 8 weeks
Vermiculite	33	40
Soilrite	80	83
Garden soil	50	60

Data were recorded after 4 and 8 weeks of transfer to planting substrates

4.1.8 Acclimatization

Plantlets were successfully hardened off according to the procedure explained in Chap. 3. Among the three planting substrates tested, 80% of the plantlets survived in soilrite (Fig. 4.19a, b, c), whereas 50 and 33% plantlets survived in garden soil and vermiculite, respectively, after 4 weeks of transplantation. Soilrite being more porous substrate holds more water than vermiculite and garden soil, and thus promoted better growth of tender roots of tissue culture (TC)-raised plants during hardening. Moreover, roots were easily penetrated in soilrite than in other planting substrates. The primary acclimatized plants on soilrite when transferred to field showed 83% survival rate after 8 weeks, whereas those on garden soil and vermiculite exhibited 60 and 40% survival, respectively (Table 4.30). The micropropagated plants showed uniform morphology and the growth indicated stability of plants free from visible abnormalities (Fig. 4.20a, b).

4.1.9 Synthetic Seeds

4.1.9.1 Effect of Alginate Concentrations on Beads Formation

The assessment of the effects of various concentrations of sodium alginate (2, 3, 4, and 5%) and $CaCl_2 \cdot 2H_2O$ (25, 50, 75, 100, and 200 mM) was prerequisite in order to standardize the preparation of characteristic beads. Calcium alginate beads containing nodal segments differed morphologically regarding texture, shape, and transparency with different concentrations of sodium alginate and $CaCl_2 \cdot 2H_2O$. An encapsulation matrix of 3% sodium alginate with 100 mM of $CaCl_2 \cdot 2H_2O$ was found most suitable for the formation of ideal beads (Fig. 4.21a) and subsequent conversion of encapsulated nodal segments into plantlets. Lower concentrations of sodium alginate [2–3% (w/v)] and $CaCl_2 \cdot 2H_2O$ (25 mM) resulted in beads without a defined shape and were too soft to handle, whereas at higher concentrations of sodium alginate (4–5%) or $CaCl_2 \cdot 2H_2O$ (200 mM), the beads were isodiametric but were hard enough to cause considerable delay in shoot emergence.

4.1.9.2 Plantlet Regeneration from Alginate-Encapsulated Nodal Segments

Encapsulated nodal segments exhibited shoot growth after 2–3 weeks of culture on different cytokinins–auxin regimes. The frequency of shoot development on different culture media varied according to medium composition (Table 4.31). MS medium supplemented with 12.5 µM BA and 1.0 µM NAA gave the maximum frequency (77%) of conversion of encapsulated nodal segments into shoots after 4 weeks (Table 4.31; Fig. 4.21b). Shoots were morphologically normal with distinct nodes and internodes.

4.1.9.3 Low Temperature Storage

After 4 weeks of storage at 4 °C, the percentage conversion of encapsulated nodal segments into shoots (4.23 ± 0.39) was 82% onto MS medium supplemented with 12.5 µM BA and 1.0 µM NAA (Table 4.32). Conversion percent of encapsulated nodal segments decreased gradually and reached

Fig. 4.19 One-month-old hardened plantlets of *Balanites aegyptiaca*

60% when the duration of storage increased to 8 weeks at 4 °C.

4.1.10 Rooting in Synthetic Seeds and Establishment of Plants in Soil

The sprouted synseeds were directly sown in autoclaved soilrite moistened with ½ MS salt solution or tap water for their conversion into plantlets. However, the sprouted synseeds did not convert into plantlets with well-developed root system. Therefore, the sprouted synseeds were then transferred to the optimized rooting medium and conversion into complete plantlet was achieved on half-strength MS medium containing 1.0 µM IBA after 4 weeks of culture (Fig. 4.21c). Thereafter, plantlets with well-developed shoot and roots were transferred to thermocol cups containing sterile soilrite and covered with polythene membrane. After 1 month, these were transferred to pots containing normal garden soil and maintained in a greenhouse. About 70% plants were successfully established in pots (Fig. 4.21d).

4.3 Clonal Fidelity in TC-Raised Plantlets Derived from Mature Nodal Explants

Fig. 4.20 Four-month-old plants acclimatized in soil under field conditions

4.2 Assessment of Physiological and Biochemical Parameters

4.2.1 Photosynthetic Pigments

Regenerated plantlets when transplanted for acclimatization showed a considerable increase in chlorophyll pigments (Chl a and Chl b) and carotenoids contents throughout the period of acclimatization. Acclimatized plantlets when exposed to external environmental conditions showed a linear increase in Chl a content from 53.4 to 84.6% against day 0 plantlet. Similarly, Chl b also showed the gradient increment of 47.8–85.7% against control plantlets (0 day). Carotenoid content increased significantly from 48% at day 0 to 84.9% at day 28 of acclimatization (Fig. 4.23).

4.2.2 Antioxidant Enzyme Activities

The changes in the antioxidant enzymatic activities of micropropagated plants during acclimatization were also studied during the study. Superoxidase dismutase (SOD) and catalase (CAT) activities increased with increasing duration of acclimatization. On the 28th day, SOD and CAT activities were the highest in the leaves than on the control day (0 day; Fig. 4.24a, b). Similarly, a significant increase in the activities of ascorbate peroxidase (APX) and glutathione reductase (GR) was observed in the plants on the 28th day of acclimatization than in the control plants (0 day; Fig. 4.25a, b).

4.3 Clonal Fidelity in TC-Raised Plantlets Derived from Mature Nodal Explants

The present study was conducted to screen TC-induced genetic variations (if any) in plantlets by employing inter-simple sequence repeat-polymerase chain reaction (ISSR-PCR) assay. Total 15 primers were initially screened and finally 5 were chosen for the present study. These 5 primers generated 59 PCR amplification products. Thus, on an average, 11.8 bands were amplified per primer. A total of 53 bands were scored for the TC-raised plantlets and 6 amplification products were specific to the outlier, thus summing up the total to 59 bands. Primer UBC 812 and 820 amplified the maximum number of 12 bands and primer UBC 819 amplified the lowest number of 9 bands (Table 4.33). In our study, all the primers

Fig. 4.21 a Sodium alginate encapsulated synthetic seeds. **b** Sprouted encapsulated synthetic seeds on MS medium containing BA (12.5 µM) and NAA (1.0 µM) after 4 weeks of incubation. **c** and **d** Synseed derived plantlets

Table 4.31 Effect of different media on conversion of encapsulated nodal segments excised from in vitro shoot cultures after 4 weeks of culture

Treatments	Conversion response (%)	Mean no. of shoots/encapsulated nodal segment
MS+BA (12.5 µM)+NAA (1.0 µM)	77	4.16 ± 0.44^a
MS+Kn (12.5 µM)+NAA (0.5 µM)	71	3.66 ± 0.33^{ab}
MS+2-iP (12.5 µM)+NAA (0.5 µM)	63	3.36 ± 0.18^{ab}
MS+TDZ (5.0 µM)	60	2.66 ± 0.44^b

MS Murashige and Skoog, *BA* benzyladenine, *NAA* a-naphthalene acetic acid, *Kn* kinetin, *2-iP* 2-isopentenyladenine, *TDZ* thidiazuron

Values represent means±SE. Means followed by the same letter within columns are not significantly different ($p=0.05$) using Duncan's multiple range test

Table 4.32 Effect of different durations on storage of encapsulated nodal segments at 4 °C on conversion into shoots

Duration (weeks)	Conversion response (%)	No. of shoots/nodal segment
0	68	3.00 ± 0.57^{abc}
1	74	3.50 ± 0.28^{ab}
2	77	3.66 ± 0.33^{ab}
4	82	4.23 ± 0.39^a
6	62	2.66 ± 0.44^{bc}
8	60	2.16 ± 0.16^c

Values represent means±SE. Means followed by the same letter within columns are not significantly different ($p=0.05$) using Duncan's multiple range test

amplified scorable bands between 100 bp to 1 kb molecular size range.

A maximum number of 14 bands were produced within the ladder size of 100–500 bp. Am-

Fig. 4.22 **a** Longitudinal section of nodal explants showing axillary bud with leaf primordia. **b** and **c** Sections showing the emergence of multiple shoot buds with leaf primordia in nodal explants. **d** Histological section showing the direct induction of adventitious shoot in root explant

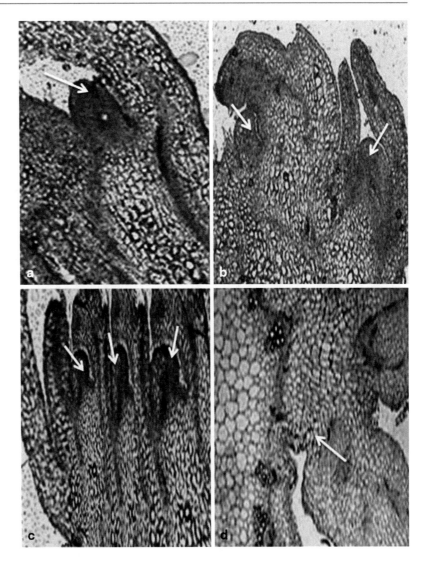

Table 4.33 Response of different UBC primers on detection of clonal stability in micropropagated plants

Primers	Primer sequence (5'–3')	Monomorphic bands in TC raised plantlets + outlier	No. of bands amplified only in outlier	Total no. of bands amplified
UBC 812	GAGAGAGAGAGAGAGA	12	2	14
UBC 814	CTCTCTCTCTCTCTCTA	10	1	11
UBC 818	CACACACACACACACAG	10	1	11
UBC 819	GTGTGTGTGTGTGTGTA	9	0	9
UBC 820	GTGTGTGTGTGTGTGTC	12	2	14
	Total number of bands produced	53	6	59

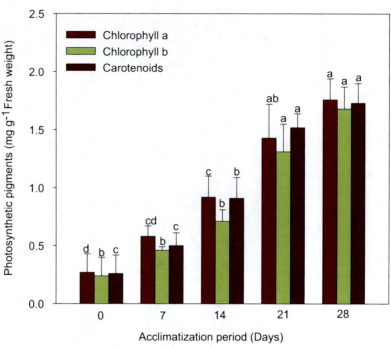

Fig. 4.23 Changes in the level of photosynthetic pigments in micropropagated plants acclimatized at photosynthetic photon flux density (PPFD) 50 µmol m^{-2} s^{-1} for 28 days. Bars represent mean ± *SE*. Bars denoted by the same letter within response variables are not significantly different ($p = 0.05$) using Duncan's multiple range test

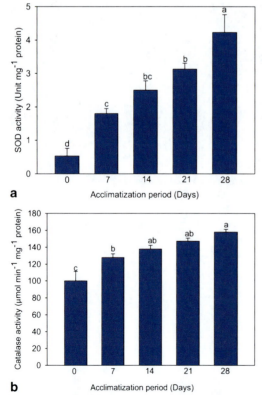

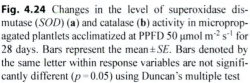

Fig. 4.24 Changes in the level of superoxidase dismutase (*SOD*) (**a**) and catalase (**b**) activity in micropropagated plantlets acclimatized at PPFD 50 µmol m^{-2} s^{-1} for 28 days. Bars represent the mean ± *SE*. Bars denoted by the same letter within response variables are not significantly different ($p = 0.05$) using Duncan's multiple test

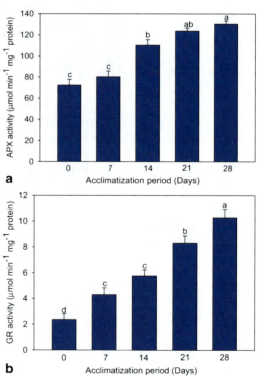

Fig. 4.25 Changes in the level of ascorbate peroxidase (*APX*) (**a**) and glutathione reductase (*GR*) (**b**) activity in micropropagated plantlets acclimatized at PPFD 50 µmol m^{-2} s^{-1} for 28 days. Bars represent the mean ± *SE*. Bars denoted by the same letter within response variables are not significantly different ($p = 0.05$) using Duncan's multiple test

Fig. 4.26 Electrophoretic gel separation of the amplification products by primers UBC 812, UBC 814 (**a**) and UBC 818 (**b**). *Lanes M1–M8* represent the TC-raised plantlets of *B. aegyptiaca*; *lane M* represents mother plant; *Lane O* represents the outlier; *Lane L* shows 1-kb ladder DNA

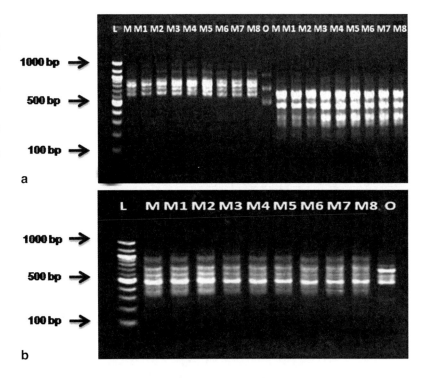

Fig. 4.27 Amplifications produced by using primers UBC 819 (**a**) and 820 (**b**). *Lane M* represents the 1-kb ladder DNA and *lanes M1–M8* represent the TC-raised plantlets. *Lane O* represents the outlier

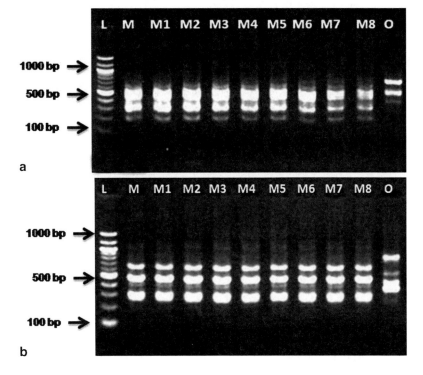

Fig. 4.28 Dendrogram illustrating similarities among regenerated plants (*Balanites 1–8*), the single donor plant (*Balanites M*), and control plant of *B. aegyptiaca* by the unweighted pair group method with arithmetic mean (UPGMA) cluster analysis using Jacard similarity coefficient based on ISSR bands data generated with five selected primers

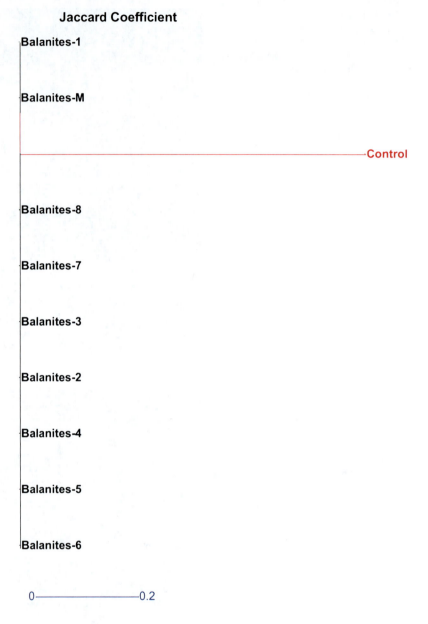

plification pattern of TC-raised *Balanites* plantlets is shown by representative gel profiles of primers UBC 812, 814, 818 (Fig. 4.26a, b), 819, and 820 (Fig. 4.27a, b). No polymorphism was detected during the ISSR analysis of TC-raised plantlets. The primer UBC 820 enabled scoring of 12 monomorphic bands for the TC-raised plantlets (lanes M–M8, Fig. 4.27b), of which two marker bands were common to the outlier (lane O, Fig. 4.27b). In the amplification profile obtained with the primer 819, nine monomorphic bands were scored for the TC-raised plants and no band was shared with the outlier. The banding profile obtained among the TC-raised progeny was completely uniform suggesting a high level of genetic fidelity among them.

Moreover, cluster analysis was performed on the basis of similarity coefficients generated from the ISSR data of the 59 scored bands. According to this ISSR analysis, all ten plants could be grouped together in a single cluster with 100% uniformity and similarity indices ranged from 0 to 0.2 among all the tested plants (Fig. 4.28).

Thus, these results confirm that the plants of *B. aegyptiaca* that multiplied through axillary method of micropropagation maintain genetic fidelity even after a prolonged period of acclimatization under ex vitro conditions.

Discussion

Abstract

Trees are major sources of food, fodder, oils, medicines, fuel, wood, fibres and timber. In many developing countries like India, the exploitation of trees is continuously increasing with population growth. Due to increased anthropogenic activities, the status of woody trees especially forest trees is greatly affected. Therefore, there is a great need to conserve forest ecosystems by agro-technology. Of the various facets of biotechnology which demand immediate application in woody trees, the reproducible protocols for regeneration of clonal plants through axillary shoot multiplication are outstanding. Biotechnological interventions for in vitro regeneration, mass micropropagation and gene transfer methods in forest tree species have been practised with success especially in the past decade. Also, propagation of woody trees through tissue culture has many advantages over conventional propagation methods like fast multiplication of the important genotypes, quick release of improved cultivars, production of disease-free plants, season-independent production of plants, germplasm conservation and facilitating their easy exchange. Keeping in mind the efficacy of micropropagation, the domino effects or results obtained during the study in *Balanites aegyptiaca* have been discussed in the light of existing literature on the subject.

5.1 Effect of Plant Growth Regulators on Different Explants

The selection of suitable explants and establishment of shoot cultures are two critical factors for the micropropagation of trees. In the present investigation, axenic cultures derived from explants of in vitro seedlings and nodal explants excised from mature trees were established. Among the different explants (cotyledonary node (CN), nodal, root, intact seedlings) tested, nodal explants derived from aseptic seedlings were found to be most excellent for attaining the maximum shoot induction and multiplication rate in *Balanites aegyptiaca*. A similar response from nodal segments excised from aseptic seedlings of woody trees was also reported to be the most effective in plant regeneration in *Syzygium cuminii* (Jain and Babbar 2000), *Terminalia chebula* (Shyamkumar et al. 2003), *Boswellia ovalifoliolata* (Chandrasekhar et al. 2005), *Feronia limonia* (Hiregoudar et al. 2005), *Pterocarpus santalinus* (Rajeswari and Paliwal 2006), *Stereospermum personatum* (Shukla et al. 2009) and *Sapindus trifoliatus* (Asthana et al. 2011).

As CN explants have already been recommended, as an excellent tissue, in several studies conducted in other tree species (Pradhan et al. 1998; Sinha et al. 2000; Anis et al. 2005), in my study also, the CN explant was found to be the second most effectual explant in obtaining a good quality of shoots for large-scale multiplication and conservation.

In *B. aegyptiaca*, a meticulous study on the regeneration potential of various explants under the influence of different plant growth regulators has been carried out, leading to the formation of a reliable protocol for its rapid multiplication. In the present investigation, 6-benzyladenine (BA), at a range of concentrations tested, was more effective in shoot induction as compared to kinetin (Kn), 2-isopentenyladenine (2-iP) and thidiazuron (TDZ), and a concentration of 12.5 μM was found to be optimum for inducing a regeneration response. This study is in accordance with the earlier reports on *B. aegyptiaca*, where BA was effective in the induction of the maximum number of multiple shoots (Siddique and Anis 2009a; Anis et al. 2010). Also, the effectiveness of BA on multiple shoot bud differentiation has been demonstrated in a number of cases (Jeong et al. 2001; Loc et al. 2005). Moreover, Zhang et al. (2010) while studying the metabolism of BA-treated mature pine plants suggested that BA causes reinvigoration of mature/old tissues and bud induction, a prerequisite for cloning of mature trees.

In this study, a higher concentration (15.0 μM) of BA did not improve any parameters and it suppressed regeneration frequency, number of shoots and shoot length. The exposure of explants to higher cytokinin concentrations during the induction phase may have led to the accumulation of cytokinins, which inhibited further shoot multiplication and growth (Malik et al. 2005).

Following the best performance of BA in multiple shoot regeneration, Kn showed a better effect than 2-iP. TDZ was the least effective of all the cytokinins tested for stimulating a high percentage of response in my study. Although multiple shoots were induced on a TDZ-containing medium, these shoots failed to elongate and were often fasciated. The formation of stunted shoots on a TDZ-supplemented medium correlates with observations in other plant species (Preece and Imel 1991; Pradhan et al. 1998; Faisal et al. 2005; Siddique et al. 2006; Siddique and Anis 2007b; Varshney and Anis 2012). The inhibition of shoot elongation may have been due to the high cytokinin activity of TDZ, and the presence of a phenyl group in TDZ may have caused fasciation (Huetteman and Preece 1993).

Experiments involving a combination of auxin and cytokinins together showed a significant increment in the rate of shoot production from each explant tried. The combination of indole-3-acetic acid (IAA) or indole-3-butyric acid (IBA) with either optimal concentrations of BA produced similar results, while the combination of α-naphthalene acetic acid (NAA) with BA was appreciably more thriving than with BA alone in terms of percent regeneration, mean number of shoots per explant and mean shoot length. An approximately twofold increase in the shoot number following the inclusion of a low concentration of NAA into a BA-supplemented medium strongly suggests an interactive effect between the cytokinin and auxin on shoot abundance. Among all combinations tested, the Murashige and Skoog (MS) medium supplemented with 12.5 μM BA and 1.0 μM NAA was found to be the most effective for shoot multiplication with maximum shoot growth in nodal explants. The shoot number, i.e. 18.00 ± 0.11, appears to be the highest as compared to the previous report by Siddique and Anis (2009a) and Anis et al. (2010), where the authors have reported a significantly fair number (11.5 ± 0.7) of shoots per explant; 7.7 ± 0.40 shoots per nodal explant excised from candidate plus tree on the BA–NAA combination, respectively.

On increasing the concentration of BA to 15.0 μM either singly or in combination with NAA to 2.5 μM, the shoot multiplication capability was adversely affected and significantly low rates of multiplication were obtained. A similar observation is substantiated by the reports where cytokinin singly or in combination with auxins beyond their optimum level exhibits a decline in the number of shoots from explants used (Ahmad et al. 2008). The optimal endogenous and exogenous cytokinin/auxin balance subsequently controlled the development and elongation of shoots.

The extensive role played by a low concentration of auxin in conjunction with cytokinin on shoot regeneration is well documented in several trees (Mercier et al. 1992; Luis et al. 1999; Yang et al. 1993; Caboni et al. 2002; Siril and Dhar 1997; Koroch et al. 2003; Shailendra et al. 2005; Ahmad and Anis 2007; Mallikarjuna and Rajendrudu 2007), and the results achieved in these plants are in accordance with the results obtained in my study.

Despite the fact that the greatest number of shoots has been obtained from nodal explants, moderately good counts of shoots were also recorded from aseptic CN explants onto medium containing 12.5 µM BA and 1.0 µM NAA and showed reasonable performance in axillary shoot regeneration in the current study. These observations corroborated the results obtained by Parveen et al. (2010) where the CN has showed better response than the other explants in *Cassia siamea*. The presence of expanded cotyledons in the CN explants might explain the reason for their better response in terms of shoot growth and development as the cell wall polysaccharides stored in them might be degraded and mobilized to the developing shoots, thereby enhancing the growth of developing shoots (Rajeswari and Paliwal 2008).

Previously, an inferior attempt on direct in vitro shoot morphogenesis from root segments cultured on B5 medium containing 0.02 mg/l NAA with 15.3 % shoot regeneration frequency was reported in *Balanites aegyptiaca* by Gour et al. (2005). On the other hand, results from the study by Varshney and Anis (2013) on culturing the roots on MS medium containing the combination of BA (5.0 µM) and NAA (1.0 µM) showed possible adventitious shoot multiplication with a favourable number of shoots per root explant in 68 % cultures. These results are in agreement with the findings of Georges et al. (1993) in *Lonicera japonica*, Shahin-uz-zaman et al. (2008) in *Azadirachta indica* and Perveen et al. (2011) in *Albizia lebbeck*, where the BA and NAA combination was best in inducing direct caulogenesis from the root culture.

The viability of any species depends on the balance between extinction and colonization. Because of the rarity of colonization, the survival of the species as a whole critically depends upon the survival of the population and germination of the seeds. The use of seeds is generally a preferable approach to conserve a species, because it helps in conserving the genetic variation in the species. If plants are multiplied from seeds, the genetic diversity of local ecotypes is maximized (Fay 1992). Each species has particular requirements for seed germination. The germination obtained on MS basal medium was found satisfactory for *Balanites aegyptiaca*, as 70 % of the seeds were germinated after 4 weeks. Observing the seed germination rate, we assume that seeds were under the influence of dormancy due to physiological and biochemical factors.

In the past, the efficacy of germinating seeds in the presence of cytokinins or a substance with cytokinin-like activity for subsequent regeneration from seedling explants has been shown in *Phaseolus vulgaris* (Malik and Saxena 1992), *Cajanus cajan* (Prakash et al. 1994), *Murraya koenigii* (Bhuyan et al. 1997), *Dendrocalamus asper* (Arya et al. 1999), *Litchi chinensis* (Das et al. 1999), *Sterculia urens* (Hussain et al. 2008) and *Pisum sativum* (Zhihui et al. 2009). However, in *Balanites aegyptiaca*, the seeds cultured on gibberellic acid (GA3)-supplemented medium elicited the morphogenic response ranging from 29 to 80 % seed germination into healthy intact seedlings.

Hence, the data from the *Balanites aegyptiaca* study provide strong substantiation that GA3 added to the medium affects the percentage response and the number of shoots regenerated per seed. The reason for this morphogenesis may be the stimulatory effects of GA3 on cell enlargement and cell division, because it is a very potent hormone. The induction of multiple shoots on seeds occurred frequently when seedlings were exposed to low levels of GA3, implying that in *Balanites aegyptiaca* organogenesis can also be induced from tissues without meristems. Furthermore, the benefits of using GA3 singly or in combination with other plant growth regulators in the culture medium for shoot multiplication have been well documented in a number of plant species (Kotsias and Roussos 2001; Farhatullah and Abbas 2007; Moshkov et al. 2008). In the best treatment of germination (culture of seeds on 2.0 µM GA3 for 4 weeks), more than four shoots were isolated from one seed. This shoot clump has produced about 15 shoots per explant when transferred to a medium containing BA (12.5 µM) and NAA (1.0 µM).

5.2 Effect of Different Media, Sucrose Concentrations and pH Levels

After the establishment of the optimized concentration of plant growth regulators for the induction of direct multiple shoots from nodal explants in this study, the experiment was also designed to evaluate the effect of different basal media on induction as well as proliferation of shoots. The MS medium was found to be the most appropriate medium for maximum shoot induction and proliferation in *Balanites aegyptiaca*. The significant and distinguishing feature of MS (1962) medium is its high nitrate, ammonium and potassium contents. Contrary to our results, Wang et al. (2005) reported that the regeneration frequencies of axillary buds were very low on the MS medium and concluded that woody plant medium (WPM) and B5 medium are effective basal media for axillary bud regeneration of *Camptotheca acuminata*. But, the success of MS medium over other basal media has been reported in *Glycine max* (Shan et al. 2005), *Holarrhena antidysenterica* (Mallikarjuna and Rajendrudu 2007), *Cassia angustifolia* (Siddique and Anis 2007), *Acacia nilotica* (Abbas et al. 2010), *Nyctanthes arbor-tristis* (Jahan et al. 2011) and *Albizzia lebbeck* (Perveen et al. 2011), which supports my observations in *Balanites aegyptiaca*.

Hazarika (2003) has documented that the supplementation of sucrose in the growth medium meets the energy demands for growth and physiological function, and authors generally use 3 % sucrose in the medium as per the recommendation of Murashige and Skoog (1962). Moreover, Langford and Wainwright (1987) found that sucrose supplied at a concentration of 3 % in the medium increased the photosynthetic ability, thereby improving the survival of plantlets. Besides serving as an energy source, it also provides the carbon precursors for structural and functional components (Marino et al. 1993). Based on the essentials, in this study too, among the considered concentrations of sucrose, 3 % sucrose in the optimal plant growth regulator regime (MS medium augmented with 12.5 µM BA and 1.0 µM NAA) was found to be the finest for in vitro multiplication and growth of shoots. Below and above 3 % sucrose, the yield of shoot number decreased and poor development with yellowing of leaves was noticed in *Balanites aegyptiaca*. The same results have been recorded in *Citrus* (Hazarika et al. 2000), *Eucomis autumnalis* (Taylor and Van Staden 2001), *Gerbera* (Ashwath and Choudhary 2002), *Camptotheca acuminata* (Wang et al. 2005), *Alocasia amazonica* (Jo et al. 2009), *Bacopa monnieri* (Naik et al. 2010) and *Tecomella undulata* (Varshney and Anis 2012).

Every species requires an optimum pH which can promote maximum shoot formation. The medium pH is an important aspect for proliferating shoots in vitro. We have investigated the effect of different pH levels in the MS medium containing 12.5 µM BA and 1.0 µM NAA, and a healthier performance in all parameters on shoot development was found at pH 5.8. Lower and higher pH levels showed low performance for the induction and proliferation of shoots. The present study suggested that shoot regeneration is affected by the changes in the media pH which can be explained by the differential uptake of nitrogen sources; the uptake of NO_3- leads to a drift towards an alkaline pH, while the uptake of NH_4+ results in a shift towards acidity. Like our results, the proliferation of shoots significantly increased when the pH of the culture medium was adjusted to 5.8 in a number of plant species such as *Calophyllum apetalum* (Nair and Seeni 2003), *Camptotheca acuminata* (Wang et al. 2005), *Mucuna pruriens* (Faisal et al. 2006a, 2006b) and *Tecomella undulata* (Varshney and Anis 2012).

5.3 Rooting and Acclimatization

Micropropagation protocols are successful only when the rate of survival of the in vitro regenerated plantlet is very high after transplantation.

This in turn depends mainly on the development of a proper root system. In this study, the strength of the MS medium appeared to be an important factor in influencing rooting efficiency as full-strength MS medium did not induce any root in the microshoots, whereas ½ MS medium induces one root per shoot. Also, the rooting methods in our study revealed that the presence of an exogenous auxin was vital for the in vitro root induction of microshoots, and IBA has been found to be the most effective auxin for in vitro rooting in Balanites shoots, followed by NAA and IAA.

The superior effects of IBA on the root development may be due to several factors such as its preferential uptake, transport and stability over other auxins and subsequent gene activation (Ludwig-Muller 2000). IBA has been reported to have a stimulatory effect on root induction in many tree species including *Morus alba* (Chand et al. 1995), *Murraya koenigii* (Bhuyan et al. 1997), *Syzygium alternifolium* (Sha Valli Khan et al. 1997), *Azadirachta indica* (Eswara et al. 1998), *Zyziphus jujuba* (Hossain et al. 2003) and *Cotinus coggygria* (Metivier et al. 2007). The poor rooting response on a medium supplemented with IAA may be due to the fact that the exogenous application of IAA is rapidly inactivated in excised plant tissue by IAA oxidase (Sembdner et al. 1980).

Though ex vitro rooting has been attempted as a means to decrease the micropropagation cost and also the time from laboratory to field (Martin 2003), in my study, this method has not provided satisfactory results as compared to the in vitro method. Similarly, the growth and survival of microcuttings rooted ex vitro is lower than from in vitro-rooted microcuttings (De Klerk 2000, 2002). The results obtained in *Balanites aegyptiaca* corroborate with the study by Diaz-Perez et al. (1995) who documented that in vitro root development usually enhances transplanting success because functioning roots can create a favourable plant water balance. Also, roots developed in vitro are believed to compensate for water loss caused by malfunctioning stomata. Improved performance and increases in dry weight of these in vitro-rooted plants may be due to extra nutrient uptake through the roots.

Acclimatization of micropropagated plants to a greenhouse or a field environment is essential because there is a difference in the micropropagation environment (e.g. high air humidity, low irradiance, low CO_2, high levels of sugars and plant growth regulators) and the greenhouse or field environment. Successful acclimatization procedures provide optimal conditions for higher survival, subsequent growth and establishment of micropropagated plants. The physiological and anatomical characteristics of micropropagated plantlets necessitate that they be gradually acclimatized to the natural environment (greenhouse or field). Techniques that more satisfactorily address the changes required for successful acclimatization necessitate lower relative humidity, higher light level, autotrophic growth and a septic environment that are distinctive circumstances of the greenhouse or field (Hazarika 2006).

Furthermore, the survival of plantlets ex vitro depends on its ability to withstand water loss and carry out photosynthesis which is enhanced by gradual acclimatization and hardening. Among the different types of planting mixtures used, an 80 % survival rate was achieved in soilrite as compared to others during the study. These observations showed consistency with the findings of Tiwari et al. (2001), Faisal et al. (2006a), Siddique and Anis (2008), Anis et al. (2010) and Varshney and Anis (2012) where the micropropagated plantlets showed the highest survival rate in soilrite and they eventually established in soil.

Additionally, during the acclimatization, the physiological change happens to be critical for the survival and re-establishment of growth of micropropagated plantlets. Moreover, acclimatization has a striking consequence on leaf chlorophyll concentration. Plantlets of *Balanites aegyptiaca* had a noticeable augmentation in the Chl a and Chl b content during the initial days of transplantation and became higher at the 28th day of ex vitro transfer. Similarly, an increase in the chlorophyll contents (Chl a and b) after transplantation has been reported by Trillas et al. (1995), Rival et al. (1997), Pospisilova et al. (1998), Van Huylenbroeck et al. (2000), Osorio et al. (2005), Guan et al. (2008), Brito et al. (2009) and Siddique and Anis (2009b).

In this study, in vitro-grown plantlets have also exhibited increased levels of carotenoids after 7 days following acclimatization. Such an upgradation in carotenoid levels is not unexpected as carotenoids are reported to be involved in protecting the photosynthetic machinery from photooxidative damage (Donnelly and Vidaver 1984; Young 1991; Van Huylenbroeck et al. 2000; Ali et al. 2005). Such an increase in the photosynthetic ability might be attributed to the improvement in chloroplast ultrastructure (Wetsztein and Sommer 1982).

The results show conformity with other studies in different plant species such as peach x almond (Trillas et al. 1995), oil palm (Rival et al. 1997), tobacco (Synkova 1997; Pospisilova et al. 1998), *Calathea* (Van Huylenbroeck et al. 1998a), carob tree (Osorio et al. 2005), *Rauvolfia tetraphylla* (Faisal and Anis 2009) and *Tylophora indica* (Faisal and Anis 2010) where an elevated level of photosynthetic competency was recorded during the final days of acclimation.

5.4 Synthetic Seeds

According to Bornman (1993), synthetic seeds may provide the only technology realistically amenable to the extensive scale-up required for the commercial production of some clones. In addition, Mathur et al. (1989) reported that the use of this technology economized upon the medium, space and time requirements. Also, a novel procedure in synthetic seed technology was reported with the use of non-embryogenic (unipolar) plant propagules. The main advantage of using vegetative propagules for the preparation of synthetic seeds would be in those cases where somatic embryogenesis is not well established or somatic embryos do not germinate into complete plantlets. In such cases, synthetic seed production from nodal segments can be used for cost-effective mass clonal propagation, potential long-term germplasm storage and delivery of tissue-cultured plants.

A successful propagation system routed through encapsulation is based on the selection of suitable plant parts as the starting plant material, the critical evaluation of factors affecting the gel matrix formation and optimization of the process of germination for plant retrieval. Encapsulation of nodal segments derived from in vitro shoot cultures was influenced by the concentration of sodium alginate and calcium chloride. In the present study, the formation of firm, uniform calcium alginate beads was achieved with 3% sodium alginate complexed with 100 mM $CaCl_2$. The sodium alginate preparation at low concentrations becomes unsuitable for encapsulation, probably because of a reduction in its gelling capacity. At higher concentrations of sodium alginate, the beads were harder, which may have suppressed the emerging shoots and roots. This differential response may be due to a synergistic effect of alginate and calcium concentration. Both sodium alginate and calcium chloride play an important role in complexation and capsule hardness (Singh et al. 2010). Ahmad and Anis (2010) reported that a 3% solution of sodium alginate upon complexation with 100 mM $CaCl_2 \cdot 2H_2O$ solution gave optimal, firm and round beads in *Vitex negundo*.

According to Kozai et al. (1991), conversion is an important factor for the success of synseed technology. In the present study, the best conversion frequency (77%) into plantlets using MS medium supplemented with 12.5 µM BA and 1.0 µM NAA could be attributed to the inclusion of MS salts and hormones into the encapsulation matrix which apparently served as a nutrient bed around the explants, facilitated growth and survival and allowed to germinate (Antonietta et al. 1999; Ara et al. 2000; Ganapathi et al. 2001). My results are in accordance with the reports of Hassan (2003), Faisal et al. (2006), Ahmad and Anis (2010) and Singh et al. (2010). In addition, the regeneration frequency was clearly influenced by storage time. With an increase in storage time to more than 4 weeks, the conversion frequency decreased considerably (60%). The decline in conversion response could be attributed to the inhibition of tissue respiration by the alginate matrix or a loss of moisture due to partial desiccation during storage as reported earlier (Danso and Ford Lloyd 2003; Faisal et al. 2006; Faisal and Anis 2007; Ahmad and Anis 2010). The observations with cold-stored encapsulated nodal segments of *Balanites aegyptiaca* at low temperature for 4 weeks are in accordance with previous studies on other species (Kinoshita and Saito 1990; Adriani et al. 2000; Tsvetkov and Hausman 2005; Faisal et al. 2006; Faisal and Anis 2007; Ahmad and Anis 2010; Singh et al. 2010).

5.5 Antioxidant Enzymes Activities

During in vitro conditions, plants grow under specific climatic conditions of high relative humidity, low CO_2 concentration and low light intensity in a culture medium with a large concentration of sugar. These special conditions result in a formation of plants with abnormal morphology, anatomy and physiology (Pospisilova et al. 1999; Estrada-Luna et al. 2001; Hazarika 2006; Pinto et al. 2011). During the transfer to ex vitro conditions, in vitro plants are exposed to light intensities higher than those used under in vitro conditions, resulting usually in photoinhibition (Carvalho and Amancio 2002; Ali et al. 2005; Osorio et al. 2010). In addition, the high differential vapour pressure between in vitro and ex vitro conditions can induce water stress. These stresses may lead to an imbalance between light energy absorption and light energy utilization in acclimatized plants and ultimately to the formation of reactive oxygen species (ROS; Ali et al. 2005; Batkova et al. 2008; Faisal and Anis 2009, 2010).

Plants possess multiple means to prevent/ minimize the deleterious effects of excess light absorption, such as metabolites that detoxify ROS and a mechanism that safely dissipates excess absorbed light (thermal energy dissipation), preventing ROS formation. However, if ROS are formed, they must be removed immediately since they can damage biomolecules such as lipids, proteins, pigments and nucleic acids (Schutzendubel and Polle 2002). As a protection against ROS, plants cells develop antioxidant enzymes that can neutralize free radicals and reduce the potential damage (Asada 2006). During the present investigation, changes in the activities of superoxide dismutase (SOD), catalase (CAT), ascorbate peroxidase (APX) and glutathione reductase (GR) have been detected. A sudden rise in the activity of SOD, CAT, APX and GR after 1 week of transplantation in plantlets suggests its role in combating oxidative stress. In accordance with my work, Van Huylenbroeck et al. (1998b, 2000) reported that micropropagated *Calathea* plantlets developed an antioxidant mechanism during acclimatization. Moreover, these authors suggest that the increase in SOD, CAT, APX and GR activities reveal a protection against photo-oxidative stress linked to photoinhibition. Additionally, an increase in CAT activity also suggests its possible role in the photorespiratory detoxification of hydrogen peroxide through the mitochondrial electron transport system (Scandalios 1990). A similar increment in the activities of these antioxidant enzymes has also been observed in micropropagated plantlets of *Phalaenopsis* (Ali et al. 2005), *Gerbera jamesonii* (Chakrabarty and Datta 2008), *Zingiber officinale* (Guan et al. 2008), *Rauvolfia tetraphylla* (Faisal and Anis 2009), *Ocimum basilicum* (Siddique and Anis 2009b), *Tylophora indica* (Faisal and Anis 2010) and *Ulmus minor* (Dias et al. 2011) against oxidative stress.

The ascorbic acid-dependent antioxidant enzymes, e.g. APX and GR, are predominantly localized in the chloroplast, which is the major site of H_2O_2 production in leaves (Foyer et al. 1997). However, the activities of antioxidant enzymes in other cell compartments are also important in ROS scavenging and suggest the formation of ROS in mitochondria and peroxisomes (Van Huylenbroeck et al. 2000; Faisal and Anis 2009). GR is considered a key enzyme responsible for maintaining the reduced form of the glutathione pool (Foyer et al. 1997). In my study, the increase in GR activity in the different stages of acclimatization suggests that GR may play an important role in scavenging H_2O_2 that is produced in the chloroplast of new leaves, under low and high light intensities during the acclimatization period.

It is clear from the present study that with the above antioxidant response under light stress, *Balanites* plantlets have the potential to scavenge the ROS generated by oxidative stress.

5.6 Assessment of Genetic Fidelity

The technique of micropropagation for the large-scale commercial production of *Balanites aegyptiaca* has been employed. However, the scaling up of any micropropagation protocol is severely hindered due to incidences of somaclonal variations, as, e.g. in the case of oil palm, where aberrant flowering patterns were observed among the regenerated plants (Matthes et al. 2001). Hence, a stringent quality check in terms of genetic similarity of the progeny becomes mandatory. Somaclonal variation mostly occurs as a response to the stress imposed on the plant under culture conditions and is manifested in the form of DNA methylations, chromosome rearrangements and point mutations (Phillips et al. 1994).

Molecular markers have come up as the most desirable tool for establishing genetic uniformity of the micropropagated plantlets. The present study documents the genetic fidelity of the tissue-cultured *B. aegyptiaca* plants as assessed by inter-simple sequence repeat (ISSR) analysis. In comparison to molecular assays such as amplified fragment length polymorphism (AFLP) and restriction fragment length polymorphism (RFLP), ISSR is cost efficient, overcomes hazards of radioactivity and requires lesser amounts of DNA (25–50 ng). Further ISSR markers have higher reproducibility than random amplification for polymorphic DNAs (random amplified polymorphic DNAs (RAPDs); Meyer et al. 1993; Fang and Roose 1997), are more informative (Nagaoka and Ogihara 1997), require no prior sequence information and hence were the choice markers for the present study. Also, the mentioned advantage of the cost efficiency associated with the ISSR assay can help in a regular genetic uniformity check of the micropropagated plantlets without adding much to the cost of tissue culture-raised plants.

The results here showed a high level of genetic similarity and detected similarity indices ranging from 0.0 to 0.2 between all of the tested plantlets. A similar conclusion was reached by the study on *Platanus acerifolia*, where a high level of similarity index ranged from 91 to 100% and only 2.88% genetic variation was detected (Huang et al. 2009). Similarly, based on the ISSR band data, a low level of somaclonal variation was deemed in the genetic stability of the micropropagated lines (Saker et al. 2006; Chandrika et al. 2010; Kumar et al. 2011).

The use of the ISSR molecular marker system to successfully assess genetic variations within the in vitro materials of a woody species provides a substantially more rapid and more effective system than the confirmation of adult phenotypic characteristics. Furthermore, my study demonstrates that axillary bud multiplication is a safe method for producing true-to-type plants. There are a number of reports in the literature which cite similar results in different plants such as *Swertia chirayita* (Joshi and Dhawan 2007), *Clerodendrum serratum* (Sharma et al. 2009), *Dendrocalamus hamiltonii* (Agnihotri et al. 2009), *Capparis decidua* (Tyagi et al. 2010), *Cymbopogon martini* (Bhattacharya et al. 2010) and *Simmondsia chinensis* (Kumar et al. 2011). Considering the importance of genetic stability in the germplasm conservation programme, the protocol of this study appears to be highly effective, as it does not allow induction of variations in *B. aegyptiaca*.

References

Abbas H, Qaiser M, Naqvi B (2010) Rapid in vitro multiplication of *Acacia nilotica* subsp. hemispherica, a critically endangered endemic taxon. Pak J Bot 42:4087–4093

Adriani M, Piccioni E, Standardi A (2000) Effects of different treatments on the conversion of 'Hayward' kiwifruit synthetic seeds to whole plants following encapsulation of in vitro derived buds. N Z J Crop Hortic Sci 28:59–67

Agnihotri RK, Mishra J, Nandi SK (2009) Improved in vitro shoot multiplication and rooting of *Dendrocalamus hamiltonii* Nees et Arn. Ex Munro: production of genetically uniform plants and field evaluation. Acta Physiol Plant 31:961–967

Ahmad N, Anis M (2007) Rapid clonal multiplication of a woody tree, *Vitex negundo* L. through axillary shoots proliferation. Agrofores Syst 71:195–200

Ahmad N, Anis M (2010) Direct plant regeneration from encapsulated nodal segments of *Vitex negundo*. Biol Plant 54:748–752

Ahmad N, Wali SA, Anis M (2008) In vitro production of true-to-type plants of *Vitex negundo* L. from nodal explants. J Hortic Sci Biotechnol 83:313–317

References

Ali MB, Hahn EJ, Paek KY (2005) Effects of light intensities on antioxidant enzymes and malondialdehyde content during short-term acclimatization on micropropagated *Phalaenopsis* plantlet. Environ Exp Bot 54:109–120

Anis M, Husain MK, Shahzad A (2005) In vitro plantlet regeneration of *Pterocarpus marsupium* Roxb., an endangered leguminous tree. Curr Sci 88:861–863

Anis M, Varshney A, Siddique I (2010) In vitro clonal propagation of *Balanites aegyptiaca* (L.) Del. Agrofores Syst 78:151–158

Antonietta GM, Emanuele P, Alvaro S (1999) Effect of encapsulation on *Citrus reticulata* Blanco. somatic embryo conversion. Plant Cell Tiss Org Cult 55:235–237

Ara H, Jaiswal U, Jaiswal VS (2000) Synthetic seed: prospects and limitations. Curr Sci 12:1438–1444

Arya S, Sharma S, Kaur R, Dev A (1999) Micropropagation of *Dendrocalamus asper* by shoot proliferation using seed. Plant Cell Rep 18:879–882

Asada K (2006) Production and scavenging of reactive oxygen species in chloroplasts and their functions. Plant Physiol 141:391–396

Ashwath CR, Choudhary ML (2002) Rapid plant regeneration from *Gerbera jamesonii* Bolus callus cultures. Acta Bot Croat 61:125–134

Asthana P, Jaiswal VS, Jaiswal U (2011) Micropropagation of *Sapindus trifoliatus* L. and assessment of genetic fidelity of micropropagated plants using RAPD analysis. Acta Physiol Plant 33:1821–1829

Batkova P, Pospisilova J, Synkova H (2008) Production of reactive oxygen species and development of antioxidative systems during in vitro growth and ex vitro transfer. Biol Plant 52:413–422

Bhattacharya S, Bandopadhyay TK, Ghosh PD (2010) High frequency clonal propagation of *Cymbopogon martinii var motia* (palmarosa) through rhizome culture and true to type assessment using ISSR marker. J Plant Biochem Biotech 19:27–274

Bhuyan AK, Pattnaik S, Chand PK (1997) Micropropagation of curry leaf tree [*Murraya koenigii* (L.) Spring.] by axillary proliferation using intact seedlings. Plant Cell Rep 16:779–782

Bornman CH (1993) Maturation of somatic embryos. In: Redenbaugh K (ed) Synseeds: applications of synthetic seeds to crop improvement. CRC Press, Boca Raton, pp 105–114

Brito G, Costa A, Coelho C, Santos C (2009) Large scale field acclimatization of Olea maderensis micropropagated plants: morphological and physiological survey. Trees 23:1019–1031

Caboni ED, Angeli S, Chiappetta A, Innocenti AM, Van Onckelen H, Damiano C (2002) Adventitious shoot regeneration from vegetative shoot apices in pear and putative role of cytokinin accumulation in the morphogenetic process. Plant Cell Tiss Org Cult 70:199–206

Carvalho LC, Amancio S (2002) Antioxidant defence system in plantlets transferred from in vitro to ex vitro: effects of increasing light intensity and CO_2 concentration. Plant Sci 162:33–40

Chakrabarty D, Dutta SK (2008) Micropropagation of Gerbera: lipid peroxidation and antioxidant enzyme activities during acclimatization process. Acta Physiol Plant 30:325–331

Chand PK, Sahoo Y, Pattnaik SK, Pattnaik SN (1995) In vitro meristem culture- an efficient ex situ conservation strategy for elite mulberry germplasm. In: Mohanty RC (ed) Environment: change and Management, Kamalraj Enterprises, New Delhi, pp 127–133

Chandrasekhar T, Hussain TM, Jayanand B (2005) In vitro micropropagation of Boswellia ovalifoliolata. Natl Cent Biotechnol Inf 60:505–507

Chandrika M, Rai RV, Thoyajaksha (2010) ISSR marker based analysis of micropropagated plantlets of *Nothapodytes foetida*. Biol Plant 54:561–565

Danso KE, Ford-Lloyd BV (2003) Encapsulation of nodal cuttings and shoot tips for storage and exchange of cassava germplasm. Plant Cell Rep 21:718–725

Das DK, Shiv Prakash N, Bhalla-Sarin N (1999) Multiple shoot induction and plant regeneration in litchi (*Litchi chinensis* Sonn.). Plant Cell Rep 18:691–695

De Klerk GJ (2000) Rooting treatment and the ex-vitrum performance of micropropagated plants. Acta Hort 530:277–288

De Klerk GJ (2002) Rooting of microcuttings: theory and practice. Vitro Cell Dev Biol-Plant 38:415–422

Dias MC, Pinto G, Santos C (2011) Acclimatization of micropropagated plantlets induces an antioxidative burst: a case study with *Ulmus minor* Mill. Photosynthetica 49:259–266

Diaz-Perez JC, Sutter EG, Shackel KA (1995) Acclimatization and subsequent gas exchange, water relations, survival and growth of microcultured apple plantlets after transplanting them in soil. Physiol Plant 95:225–232

Donnelly DJ, Vidaver WE (1984) Pigment content and gas exchange of red raspberry transferred in vitro and ex vitro. J Am Soc Hort Sci 109:177–181

Estrada-Luna AA, Davies FT, Egilla JN (2001) Physiological changes and growth of micropropagated chile ancho pepper plantlets during acclimatization and post-acclimatization. Plant Cell Tiss Org Cult 66:17–24

Eswara JP, Stuchbury T, Allan EJ (1998) A standard procedure for the micropropagation of the neem tree (*Azadirachta indica* A. Juss). Plant Cell Rep 17:215–219

Faisal M, Ahmad N, Anis M (2005) Shoot multiplication in *Rauvolfia tetraphylla* L. using thidiazuron. Plant Cell Tiss Org Cult 80:187–190

Faisal M, Ahmad N, Anis M (2006) In vitro plant regeneration from alginate-encapsulated microcuttings of *Rauvolfia tetraphylla* L. World J Agri Sci 1:1–6

Faisal M, Anis M (2007) Regeneration of plants from alginate encapsulated shoots of *Tylophora indica* (Burm. f.) Merrill, an endangered medicinal plant. J Hortic Sci Biotechnol 82:351–354

Faisal M, Anis M (2009) Changes in photosynthetic activity, pigment composition, electrolyte leakage, lipid peroxidation and antioxidant enzymes during ex vitro establishment of micropropagated *Rauvolfia tetraphylla* plantlets. Plant Cell Tiss Org Cult 99:125–132

Faisal M, Anis M (2010) Effect of light irradiations on photosynthetic machinery and antioxidative enzymes during ex vitro acclimatization of *Tylophora indica* plantlets. J Plant Interact 5:21–27

Faisal M, Siddique I, Anis M (2006) a) In vitro rapid regeneration of plantlets from nodal explants of *Mucuna pruriens*- a valuable medicinal plant. Ann Appl Biol 148:1–6

Faisal M, Siddique I, Anis M (2006) b) An efficient plant regeneration system for *Mucuna pruriens* L. (DC) using cotyledonary node explants. Vitro Cell Dev Biol Plant 42:59–64

Fang DQ, Roose ML (1997) Identification of closely related citrus cultivars with inter-simple sequence repeat markers. Theor Appl Genet 95:408–417

Farhatullah AZ, Abbas SJ (2007) In vitro effects of gibberellic acid on morphogenesis of potato explants. Int J Agri Biol 9:181–182

Fay MF (1992) Conservation of rare and endangered plants using in vitro methods. Vitro Cell Dev Biol-Plant 28:1–4

Foyer CH, Lopez-Delgado H, Dat JF, Scott IM (1997) Hydrogen peroxide and glutathione-associated mechanisms of acclimatory stress tolerance and signaling. Physiol Plant 100:241–254

Ganapathi TR, Srinivas I, Suprasanna P, Bapat VA (2001) Regeneration of plants from alginated-encapsulated somatic embryos of banana cv. Rasthali (*Musa* spp. AAB group). Biol Plant 37:178–181

Georges D, Chemienx JC, Ochatt SD (1993) Plant regeneration from aged callus of woody ornamental species *Lonicera japonica* cv. Hell's prolific. Plant Cell Rep 13:91–94

Gour VS, Emmanuel CJSK, Kant T (2005) Direct in vitro shoot morphogenesis in desert date- *Balanites aegyptiaca* (L.) Del. from root segments. In: Tewari VP, Srivastava RL (eds) Multipurpose trees in the tropics: management and improvement strategies, India Scientific Publishers, Jodhpur, pp 701–704

Guan QZ, Guo YH, Sui XL, Li W, Zhang ZX (2008) Changes in photosynthetic capacity and oxidant enzymatic systems in micropropagated *Zingiber Officinale* plantlets during their acclimation. Photosynthetica 46:193–201

Hassan NS (2003) In vitro propagation of Jojoba (*Simmondsia chinensis* L.) through alginate-encapsulated shoot apical and axillary buds. Int J Agri Biol 5:513–516

Hazarika BN (2003) Acclimatization of tissue cultured plants. Curr Sci 85:1705–1712

Hazarika BN (2006) Morpho-physiological disorders in in vitro culture of plants. Sci Hortic 108:105–120

Hazarika BN, Parthasarathy VA, Nagaraju V, Bhowmik G (2000) Sucrose induced biochemical changes in in vitro microshoots of *Citrus* species. Indian J Hortic 57:27–31

Hiregoudar LV, Kumar HGA, Murthy HN (2005) In vitro culture of *Feronia limonia* (L.) Swingle from hypocotyls and intermodal explants. Biol Plant 49:41–45

Hossain SN, Munshi MK, Islam MR, Hakim L, Hossain M (2003) In vitro propagation of Plum (*Zyziphus jujuba* Lam.). Plant Cell Tiss Org Cult 13:81–84

Huang WJ, Ning GG, Liu GF, Bao MZ (2009) Determination of genetic stability of long-term micropropagated plantlets of *Platanus acerifolia* using ISSR markers. Biol Plant 53:159–163

Huetteman CA, Preece JE (1993) Thidiazuron: a potent cytokinin for woody plant tissue culture. Plant Cell Tiss Org Cult 33:105–119

Hussain TM, Chandrasekhar T, Gopal GR (2008) Micropropagation of *Sterculia urens* Roxb., an endangered tree species from intact seedlings. Afri J Biotechnol 7:095–101

Jahan AA, Anis M, Aref MI (2011) Assessment of factors affecting micropropagation and ex vitro acclimatization of *Nyctanthes arbor-tristis* L. Acta Biol Hungarica 62:45–56

Jain N, Babbar SB (2000) Recurrent production of plants of black plum. *Syzygium cumini* (L.) Skeels, a myrtaceous fruit tree, from in vitro cultured seedling explants. Plant Cell Rep 19:519–524

Jeong JH, Murthy HN, Paek KY (2001) High frequency adventitious shoot induction and plant regeneration from leaves of statice. Plant Cell Tiss Org Cult 65:123–128

Jo EA, Tewari RK, Hahn EJ, Paek KY (2009) In vitro sucrose concentration affects growth and acclimatization of *Alocasia amazonica* plantlets. Plant Cell Tiss Org Cult 96:307–315

Joshi P, Dhawan V (2007) Assessment of genetic fidelity of micropropagated *Swerita chiraytia* plantlets by ISSR marker assay. Biol Plant 51:22–26

Kinoshita I, Saito A (1990) Propagation of Japanese white birch by encapsulated axillary buds.1. Regeneration of plants under aseptic conditions. J Jap Fores Soc 72:166–170

Koroch K, Juliani HP, Kapteyn J, Simon JE (2003) In vitro regeneration of *Echinacea purpurea* from leaf explants. Plant Cell Tiss Org Cult 69:79–83

Kotsias D, Roussos PA (2001) An investigation on the effect of different plant growth regulating compounds in in vitro shoot tip and node culture of lemon seedlings. Sci Hortic 89:115–128

Kozai T, Ting KC, Aitken-Christie J (1991) Consideration for automation of micropropagation systems. Transactions ASAE 35:503–517

Kumar S, Mangal M, Dhawan AK, Singh N (2011) Assessment of genetic fidelity of micropropagated plants of *Simmondsia chinensis* (Link.) Schneider using RAPD and ISSR markers. Acta Physiol Plant 33:2541–2545

Langford PJ, Wainwright H (1987) Effects of sucrose concentration on the photosynthetic ability of rose shoots in vitro. Ann Bot 60:633–640

Loc HN, Due TD, Kwon HT, Yang SM (2005) Micropropagation of zedoary (*Curcuma zedoaria* Roscoe): a valuable medicinal plant. Plant Cell Tiss Org Cult 81:119–122

Ludwig-Muller J (2000) Indole-3-butyric acid in plant growth and development. Plant Growth Regul 32:219–230

Luis PBC, Adriane CMGM, Silvica BRCC, Ana Christina MB (1999) Plant regeneration from seedling explants of *Eucalyptus grandis* and *Eucalyptus urophylla*. Plant Cell Tiss Org Cult 56:17–23

Malik KA, Saxena PK (1992) Somatic embryogenesis and shoot regeneration from intact seedlings of *Phaseolus actuifolius* A., *P.aureus* (L.) Wilczek, *P. concineus* L., *P. wrightii*. Plant Cell Rep 11:163–168

Malik SK, Chaudhury R, Kalia RK (2005) Rapid in vitro multiplication and conservation of *Garcinia indica*: a tropical medicinal tree species. Sci Hortic 106:539–553

Mallikarjuna K, Rajendurdu G (2007) High frequency in vitro propagation of *Holarrhena antidysentrica* from nodal buds of mature tree. Biol Plant 51:525–529

Marino G, Bertazza G, Magnanini E, Alton AD (1993) Comparative effects of sorbitol and sucrose as main carbon energy sources in micropropagation of apricot. Plant Cell Tiss Org Cult 34:235–244

Martin KP (2003) Rapid in vitro multiplication and ex vitro rooting of *Rotula aquatica* Lour., a rare rhoephytic woody medicinal plant. Plant Cell Rep 21:415–420

Mathur J, Ahuja PS, Lal N, Mathur AK (1989) Propagation of *Valeriana wallichi* DC using encapsulated apical and axial shoot buds. Plant Sci 60:111–116

Matthes M, Singh R, Cheah SC, Karp A (2001) Variation in oil palm *Elaeis guineensis* (Jacq.) tissue culture-derived regenerants revealed by AFLPs with methylation-sensitive enzymes. Theor Appl Genet 102:971–979

Mercier H, Vieira CCJ, Figueredo-Ribeiro RCL (1992) Tissue culture and plant propagation of *Gomphrena officinalis*, a Brazilian medicinal plant. Plant Cell Tiss Org Cult 28:249–254

Metivier PSR, Edward CY, Patel KR, Thorpe TA (2007) In vitro rooting of microshoots of *Cotinus coggygria* Mill, a woody ornamental plant. Vitro Cell Dev Biol-Plant 43:119–123

Meyer W, Michell TG, Freedman EZ, Vilgalys R (1993) Hybridization probes for conventional DNA fingerprinting used as single primers in polymerase chain reaction to distinguish strains of *Cryptococcus neoformans*. J Clin Biol 31:2274–2280

Moshkov IE, Novikova GV, Hall MA, George EF (2008) Plant growth regulators III: gibberellins, ethylene, abscisic acid, their analogues and inhibitors; miscellaneous compounds. In: George EF, Hall MA, De Klerk GJ (eds) Plant propagation by tissue culture. Springer, The Netherlands, pp 227–282

Murashige T, Skoog F (1962) A revised medium for rapid growth and bioassays with tobacco tissue cultures. Physiol Plant 15:473–497

Nagaoka T, Ogihara Y (1997) Applicability of inter-simple sequence repeat polymorphism in wheat for use as DNA markers in comparision to RFLP and RAPD markers. Theor Appl Genet 94:597–602

Naik PM, Manohar SH, Praveen N, Murthy HN (2010) Effects of sucrose and pH levels on in vitro shoot regeneration from leaf explants of *Bacopa monnieri* and accumulation of bacoside A in regenerated shoots. Plant Cell Tiss Org Cult 100:235–239

Nair LG, Seeni S (2003) In vitro multiplication of *Calophyllum apetalum* (Clusiaceae), an endemic medicinal tree of the Western Ghats. Plant Cell Tiss Org Cult 75:169–174

Osorio ML, Goncalves S, Osorio J, Romano A (2005) Effects of CO_2 concentration on acclimatization and physiological responses of two cultivars of carob tree. Biol Plant 49:161–167

Osorio ML, Osorio J, Romano A (2010) Chlorophyll fluorescence in micropropagated *Rhododendron ponticum* subsp. Baeticum plants in response to different irradiances. Biol Plant 54:415–422

Parveen S, Shahzad A, Saema S (2010) In vitro plant regeneration system for *Cassia siamea* Lam., a leguminous tree of economic importance. Agrofores Syst 80:109–116

Perveen S, Varshney A, Anis M, Aref IM (2011) Influence of cytokinins, basal media and pH on adventitious shoot regeneration from excised root cultures of *Albizia lebbeck*. J Fores Res 22:47–52

Phillips RL, Kaeppler SM, Olhoft P (1994) Genetic instability of plant tissue cultures: breakdown of normal controls. Proc Nat Acad Sci USA 91:5222–5226

Pinto G, Silva S, Loureiro J, Costa A, Dias MC, Araujo C, Neves L, Santos C (2011) Acclimatization of secondary somatic embryos derived plants of *Eucalyptus globules* Labill.: an ultrastructural approach. Trees 25:383–292

Pospisilova J, Ticha I, Kadlecek P, Haisel D, Plzakova S (1999) Acclimatization of micropropagated plants to ex vitro conditions. Biol Plant 42:481–497

Pospisilova J, Wilhelmova N, Synkova H, Catsky J, Krebs D, Ticha I, Hanackova B, Snopel J (1998) Acclimatization of tobacco plantlets to ex vitro conditions as affected by application of abscissic acid. J Exp Bot 49:863–869

Pradhan C, Kar S, Pattnaik S, Chand PK (1998) Propagation of *Dalbergia sissoo* Roxb. through in vitro shoot proliferation from cotyledonary nodes. Plant Cell Rep 18:122–126

Prakash SN, Deepak P, Neera BS (1994) Regeneration of pigeon pea (*Cajanus cajan* L.) from cotyledonary node via multiple shoot formation. Plant Cell Rep 13:624–627

Preece JE, Imel MR (1991) Plant regeneration from leaf explants of Rhododendron 'PJM Hybrids'. Sci Hortic 48:159–170

Rajeswari V, Paliwal K (2006) In vitro shoot multiplication and ex vitro rooting of Pterocarpus santalinus L. an endangered leguminous tree. In: 11th IAPTC & B congress on biotechnology and sustainable agriculture 2006 and beyond, Beijing, China. Abstract 1535:13–18

Rajeswari V, Paliwal K (2008) In vitro plant regeneration of red sanders (*Pterocarpus santalinus* L. f.) from cotyledonary nodes. Indian J Biotechnol 7:541–546

Rival A, Beule T, Lavengne D, Nato A, Havaux M, Puard M (1997) Development of photosynthetic characteristics in oil palm during in vitro micropropagation. J Plant Physiol 150:520–527

Saker MM, Adawy SS, Mohamed AA, El-Itriby HA (2006) Monitoring of cultivar identity in tissue culture-derived date palms using RAPD and AFLP analysis. Biol Plant 50:198–204

Scandalios G (1990) Response to plant antioxidant defense genes to environmental stress. Adv Genet 28:1–41

Schutzendubel A, Polle A (2002) Plant responses to abiotic stresses: heavy metal-induced oxidative stress and protection by mycorrhization. J Exp Bot 53:351–1365

Sembdner G, Gross D, Liebisch HW, Schneider G (1980) Biosynthesis and metabolism of plant hormones. In: Macmillan J (ed) Encyclopedia of plant physiology (new series), vol 9. Springer, Berlin Heidelberg New York, pp 281–444

Sha Valli Khan PS, Prakash E, Rao KR (1997) In vitro propagation of an endemic fruit tree *Syzygium alternifolium* (Wight) Walp. Plant Cell Rep 16:325–328

Shahin-uz-zaman M, Ashrafuzzaman H, Shahidul Haque M, Lutfun LN (2008) In vitro clonal propagation of the neem tree (*Azadirachta indica* A. Juss). Afri J Biotechnol 7:386–391

Shailendra V, Joshi N, Tak N, Purohit SD (2005) In vitro adventitious shoot bud differentiation and plantlet regeneration in *Feronia limonia* L. (Swingle). Vitro Cell Dev Biol-Plant 41:296–302

Shan Z, Raemakers K, Tzitzikas EN, Ma Z, Visser RGF (2005) Development of a highly efficient, repetitive system of organogenesis in soybean [*Glycine max* (L.) Merr]. Plant Cell Rep 24:507–512

Sharma M, Rai SK, Purshottam DK, Jain M, Chakrabarty D, Awasthi A, Nair KN, Kumar A (2009) In vitro clonal propagation of *Clerodendrum serratum* (Linn.) Moon (barangi): a rare and threatened medicinal plant. Acta Physiol Plant 31:379–383

Shukla S, Shukla SK, Mishra SK (2009) In vitro plant regeneration from seedling explants of *Stereospermum personatum* D.C.: a medicinal tree. Trees 23:409–413

Shyamkumar B, Anjaneyulu C, Giri CC (2003) Multiple shoot induction from cotyledonary node explants of *Terminalia chebula* Retz: a tree of medicinal importance. Biol Plant 47:585–588

Siddique I, Anis M (2007) High frequency shoot regeneration and plantlet formation in *Cassia angustifolia* (Vahl.) using thidiazuron. Med Aroma Plant Sci Biotechnol 1:282–284

Siddique I, Anis M (2008) An improved plant regeneration system and ex vitro acclimatization of *Ocimum basilicum* L. Acta Physiol Plant 30:493–499

Siddique I, Anis M (2009a) Direct plant regeneration from nodal explants of *Balanites aegyptiaca* L. (Del.): a valuable medicinal tree. New Fores 37:53–62

Siddique I, Anis M (2009b) Morphogenic response of the alginated encapsulated nodal segments and antioxidative enzymes analysis during acclimatization of *Ocimum basilicum* L. J Crop Sci Biotechnol 12:233–238

Siddique I, Anis M, Jahan AA (2006) Rapid multiplication of *Nyctanthes arbor-tristis* L. through in vitro axillary shoot proliferation. World J Agric Sci 2:188–192

Singh SK, Rai MK, Asthana P, Sahoo L (2010) Alginate-encapsulation of nodal segments for propagation, short-term conservation and germplasm exchange and distribution of *Eclipta alba* (L.). Acta Physiol Plant 32:607–610

Sinha RK, Majumdar K, Sinha S (2000) In vitro differentiation and plant regeneration of *Albizia chinensis* (OBS.) MERR. Vitro Cell Dev Biol-Plant 36:370–373

Siril EA, Dhar U (1997) Micropropagation of mature Chinese tallow tree (*Sapium sebiferum* Roxb.). Plant Cell Rep 16:637–640

Synkova H (1997) Sucrose affects the photosynthetic apparatus and the acclimatization of transgenic tobacco to ex vitro culture. Photosynthetica 33:403–412

Taylor JLS, Van Staden J (2001) The effect of nitrogen and sucrose concentrations on the growth of *Eucomis autumnalis* (Mill) Chitt. plantlets in vitro, and on subsequent anti-inflammatory activity in extracts prepared from the plantlets. Plant Growth Regul 34:49–56

Tiwari V, Tiwari KN, Singh BD (2001) Comparative studies of cytokinins on in vitro propagation of *Bacopa monnieri*. Plant Cell Tiss Org Cult 66:9–16

Trillas MI, Serret MD, Jorba J, Araus JL (1995) Leaf chlorophyll fluorescence changes during acclimatization of the root stock GF 667 (peach x almond) and propagation of *Gardenia jasminoides* E. In: Carre F, Chagvardieff P (eds) Ecophysiology and Photosynthetic in vitro cultures. CEA, Saint-Paul-lez-Durance. pp 161–168

Tsvetkov I, Hausman JE (2005) In vitro regeneration from alginate encapsulated microcuttings of *Quercus* sp. Sci Hort 103:503–507

Tyagi P, Khanduja S, Kothari SL (2010) In vitro culture of *Capparis decidua* and assessment of clonal fidelity of the regenerated plants. Biol Plant 54:126–130

Van Huylenbroeck JM, Piqueras A, Debergh PC (1998a) Photosynthesis and carbon metabolism in leaves formed prior to and during ex vitro acclimatization of micropropagated plants. Plant Sci 134:21–30

Van Huylenbroeck JM, Piqueras A, Debergh PC (2000) The evolution of photosynthesis capacity and the antioxidant enzymatic system during acclimatization of micropropagated Calathea plants. Plant Sci 155:59–66

Van Huylenbroeck JM, Van Laere IMB, Piqueras A, Debergh PC, Buneo P (1998b) Time course of Catalase and Superoxide dismutase during acclimatization and growth of micropropagated *Calathea* and *Spathiphyllum* plants. Plant Growth Reg 26:7–14

Varshney A, Anis M (2012) Improvement of shoot morphogenesis in vitro and assessment of changes of the activity of antioxidant enzymes during acclimation of micropropagated plants of Desert Teak. Acta Physiol Plant 34:859–867

References

Varshney A, Anis M (2013) Direct plantlet regeneration from segments of root of *Balanites aegyptiaca* Del. (L.)- a biofuel arid tree. Int J Pharm Bio Sci 4:987–999

Wang HM, Zu YG, Dong FL, Zhao XJ (2005) Assessment of factors affecting in vitro shoot regeneration from axillary bud explants of *Camptotheca acuminata*. J Fores Res 16:52–54

Wetzstein HY, Sommer HE (1982) Leaf anatomy of tissue cultured *Liquidambar styraciflua* (Hamammelidaceae) during acclimatization. Am J Bot 69:1579–1586

Yang JC, Tsay JY, Chung JD, Chen ZZ (1993) In vitro clonal propagation and cell suspension culture of *Gmelina arborea* R. Bull Taiwan Fores Res Inst 8:1–9

Young AJ (1991) The protective role of carotenoids in higher plants. Physiol Plant 83:702–708

Zhang H, Horgan KJ, Reynolds PH, Jameson PE (2010) 6-Benzyladenine metabolism during reinvigoration of mature *Pinus radiata* buds in vitro. Tree Physiol 30:514–526

Zhihui S, Tzitzikas M, Raemakers K, Zhengqiang M, Visser R (2009) Effect of TDZ on plant regeneration from mature seeds in pea (*Pisum sativum*). Vitro Cell Dev Biol- Plant 45:776–782

Summary and Conclusions

> **Abstract**
>
> *Balanites aegyptiaca* (L.) Del. is an important tree in the semi-arid ecosystem with beneficial attributes. The fruits and roots of this plant contain diosgenin, which can be used in pharmaceutical industry for the production of oral contraceptives and steroids. This chapter summarizes the enormous amount of studies undertaken in different aspects of micropropagation and plant biotechnology.

6.1 Summary

An in vitro rapid propagation system has been developed for *B. aegyptiaca* (L.) Del., a spinescent semi-arid tree with a multitude of potential uses. The fruits and roots contain a valuable secondary metabolite 'Diosgenin', which is used in the production of steroidal hormones. The seeds that were germinated on full-strength Murashige and Skoog (MS) medium served as source for the explants, and the morphogenetic potential of each explant was tested on MS medium with various concentrations (1.0–15.0 µM) of 6-benzyladenine (BA), kinetin (Kn) and thidiazuron (TDZ) alone or in combination with different concentrations (0.5–2.5 µM) of auxins [α-naphthalene acetic acid (NAA), indole-3-butyric acid (IBA) and indole-3-acetic acid]. Among the different explants (mature nodal segments and seedling-derived explants) that were tried, nodal explants from 4-week-old axenic seedlings yielded an optimal shoot regeneration frequency (80 %), maximum number (10.06 ± 0.86) of shoots and mean shoot length (4.56 ± 0.29 cm) on MS medium supplemented with 12.5 µM BA after 4 weeks of culture initiation. Further, the most excellent response in shoot multiplication and elongation was recorded when the shoot cultures were transferred into medium augmented with 12.5 µM BA and 1.0 µM NAA, and the highest number, i.e. 18.00 ± 0.11, of shoots per nodal explant with mean shoot length of 5.80 ± 0.23 cm were obtained on the same medium after 8 weeks in 85 % culture.

Half-strength MS medium supplemented with IBA gave the best result for rooting. The maximum frequency of root formation (80 %), number of roots per shoot (8.10 ± 0.60) and root length (4.3 ± 0.37 cm) were obtained on half-strength MS medium containing 1.0 µM IBA after 4 weeks. The micropropagated shoots with well-developed roots were acclimatized in soilrite and successfully transplanted to soil with 83 % survival rate.

As the acclimatization of micropropagated plantlets continued from 0 to 28 days, a linear increment both in the contents of photosynthetic pigments (Chl a, Chl b and carotenoids) and in the activities of antioxidant enzymes (superoxide dismutase, SOD; Catalase (CAT); Ascorbate Peroxidase (APX) and GR) was also ascertained.

After standardizing the reliable protocol for micropropagation, the influence of different basal media, sucrose concentrations and pH values was also assessed on the shoot morphogenesis from seedling-derived nodal explants. The MS medium was found to be the best basal medium followed by woody plant medium (WPM) and Gamborg (B_5) media containing BA (12.5 µM) and NAA (1.0 µM) after 8 weeks of culture. In the same plant growth regulator regime, medium amended with 3% sucrose at pH 5.8 showed the finest performance on shoot multiplication and proliferation after 8 weeks of incubation.

Nodal segments obtained from the in vitro-proliferated shoots were encapsulated in calcium alginate beads, and the best gel complexation was achieved using 3% sodium alginate and 100 mM $CaCl_2.2H_2O$. The maximum percent response (77%) for conversion of encapsulated nodal segments into plantlets was obtained on MS medium containing BA (12.5 µM) and NAA (1.0 µM). The encapsulated nodal segments could be stored at a low temperature (4 °C) up to 4 weeks with a survival frequency of 82%. The well-developed plantlets regenerated from encapsulated nodal segments were hardened-off successfully with 70% survival frequency.

Inter-simple sequence repeat (ISSR) markers were used to assess the genetic stability of the micropropagated plantlets. The micropropagated plantlets were chosen from a clonal collection of shoots that originated from mature nodal explants (mother plant). Based on the ISSR band data, similarity indices between the plantlets ranged from 0 to 0.2. These similarity indices were used to construct an unweighted pair group method with arithmetic mean (UPGMA) dendrogram and demonstrated that all the selected plantlets grouped together in one major cluster had a similarity level of 100%. Thus, the amplification products were monomorphic and the plantlets did not show any detectable variation either in morphological or in growth characteristics when compared with the mother plant.

6.2 Conclusions

Based on the research findings, we conclude that the accomplishment of work in establishing the plant regeneration system by tissue culture and acclimatization of regenerated plantlets to soil in woody species is a prerequisite for biotechnological approaches to plant micropropagation and improvement. The experiments conducted on *B. aegytiaca* proved that plantlet formation from nodal segment is practicable and can provide a wide possibility for conservation and an effective way for the propagation of other tropical tree species. The use of the ISSR molecular marker system to successfully assess genetic variations within the in vitro materials of a woody species provides a substantially more rapid and more effective system than confirmation of adult phenotypic characteristics. Such clonal plant materials are therefore suitable as the basis of an imperative programme of *Agrobacterium*-mediated genetic transformation.

Also, the synthetic seed technology described in Chap. 4 (Results) provides an alternative method for the propagation of this important medicinal plant. Successful plant retrieval from encapsulated nodal segments following at low temperature indicates that the method described here could be potentially used to preserve desirable elite genotypes of *B. aegyptiaca* for a short period. This could also facilitate transport of encapsulated nodal segments to laboratories and extension centres at distant places.

Chapter 4 (Results) also demonstrated that the micropropagated plantlets developed significantly higher levels of antioxidant enzymes activities, which play an important role for better environmental adaptation of transplanted plantlets from in vitro conditions. Since the plants get affected with increasing light intensity due to changing environmental conditions, this study provides important information for the development of medicinally and economically important plants.

Thus, the process can be exploited for catering enough raw material to various pharmaceutical industries as well as for providing large number of cloned plants of *B. aegyptiaca* for its rehabilitation in natural habitat for conservation and sustainable utilization.

Printed by Publishers' Graphics LLC